Introduction to Online Control

This tutorial guide introduces online nonstochastic control, an emerging paradigm in control of dynamical systems and differentiable reinforcement learning that applies techniques from online convex optimization and convex relaxations to obtain new methods with provable guarantees for classical settings in optimal and robust control. In optimal control, robust control, and other control methodologies that assume stochastic noise, the goal is to perform comparably to an offline optimal strategy.

In online control, both cost functions and perturbations from the assumed dynamical model are chosen by an adversary. Thus, the optimal policy is not defined a priori, and the goal is to attain low regret against the best policy in hindsight from a benchmark class of policies. The resulting methods are based on iterative mathematical optimization algorithms and are accompanied by finite-time regret and computational complexity guarantees.

This book is ideal for graduate students and researchers interested in bridging classical control theory and modern machine learning.

ELAD HAZAN is Professor of Computer Science at Princeton University. His research focuses on the design and analysis of algorithms for basic problems in machine learning and optimization. He is a pioneer of online nonstochastic control theory.

KARAN SINGH is Assistant Professor of Operations Research at Carnegie Mellon University and has previously worked at Google Brain and Microsoft Research. He works on the foundations of machine learning, control, and reinforcement learning.

"We are in a golden age for control and decision making. A proliferation of new applications including self-driving vehicles, humanoid robots, and artificially intelligent drones opens a new set of challenges for control theory to address. Hazan and Singh have written the definitive book on the New Control Theory – non-stochastic control. The phrase 'a paradigm shift' has become cliche from overuse, but here it is truly well deserved; the authors have revisited the foundations by focusing on building controllers that perform nearly as well as if they knew future disturbances in advance, rather than relying on probabilistic or worst-case models. The non-stochastic control approach has extended one of the most profound ideas in mathematics of the 20th century, online (no-regret) learning, to master sequential decision making with continuous actions. This leads to high performance in benign environments and resilience in adversarial ones. The book, authored by pioneers in the field, presents both foundational concepts and the latest research, making it an invaluable resource."

– *Drew Bagnell, Carnegie Mellon University and Aurora*

"As someone who has worked extensively on learning theory and online learning, and later applied these ideas in domains such as autonomous driving and humanoid robotics, I find this book both timely and inspiring. It introduces a regret-minimization framework for control that draws on the elegance and power of online learning. Traditional control theory often models noise either as stochastic–sometimes unrealistically optimistic–or adversarial–often overly conservative. This book charts a new path by asking a deeper question: while we cannot predict noise, can we perform nearly as well as if we could? The answer, developed here, is a novel and exciting paradigm that bridges learning theory and control, and I believe it will have a lasting impact on both research and practice."

– *Shai Shalev-Shwartz, The Hebrew University of Jerusalem*

Introduction to Online Control

ELAD HAZAN
Princeton University

KARAN SINGH
Carnegie Mellon University

Shaftesbury Road, Cambridge CB2 8EA, United Kingdom

One Liberty Plaza, 20th Floor, New York, NY 10006, USA

477 Williamstown Road, Port Melbourne, VIC 3207, Australia

314–321, 3rd Floor, Plot 3, Splendor Forum, Jasola District Centre,
New Delhi – 110025, India

103 Penang Road, #05–06/07, Visioncrest Commercial, Singapore 238467

Cambridge University Press is part of Cambridge University Press & Assessment,
a department of the University of Cambridge.

We share the University's mission to contribute to society through the pursuit of
education, learning and research at the highest international levels of excellence.

www.cambridge.org
Information on this title: www.cambridge.org/9781009499668
DOI: 10.1017/9781009499705

© Elad Hazan and Karan Singh 2026

This publication is in copyright. Subject to statutory exception and to the provisions
of relevant collective licensing agreements, no reproduction of any part may take place
without the written permission of Cambridge University Press & Assessment.

When citing this work, please include a reference to the DOI 10.1017/9781009499705

First published 2026

Cover image: Elad Hazan

A catalogue record for this publication is available from the British Library

Library of Congress Cataloging-in-Publication Data
Names: Hazan, Elad, 1975– author | Singh, Karan (Assistant professor of operations
research) author
Title: Introduction to online control / Elad Hazan, Karan Singh.
Description: Cambridge ; New York, NY : Cambridge University Press, 2026. | Includes
bibliographical references.
Identifiers: LCCN 2025029698 (print) | LCCN 2025029699 (ebook) | ISBN
9781009499668 hardback | ISBN 9781009499699 paperback | ISBN 9781009499705 epub
Subjects: LCSH: Control theory–Mathematical models | Online algorithms | Mathematical
optimization | Convex domains
Classification: LCC QA402.3 .H365 2025 (print) | LCC QA402.3 (ebook)
LC record available at https://lccn.loc.gov/2025029698
LC ebook record available at https://lccn.loc.gov/2025029699

ISBN 978-1-009-49966-8 Hardback

Cambridge University Press & Assessment has no responsibility for the persistence
or accuracy of URLs for external or third-party internet websites referred to in this
publication and does not guarantee that any content on such websites is, or will
remain, accurate or appropriate.

For EU product safety concerns, contact us at Calle de José Abascal, 56, 1°, 28003 Madrid,
Spain, or email eugpsr@cambridge.org

To my family: Dana, Hadar, Yoav, Oded, and Deluca,
—EH

To my parents – A & V.
—KS

Contents

Preface		page xi
Acknowledgments		xii
List of Symbols		xiv

	Part I	**Background in Control and Reinforcement Learning**	1
1	**Introduction**		3
	1.1	What Is This Book About?	3
	1.2	The Origins of Control	4
	1.3	Formalization and Examples of a Control Problem	5
	1.4	Simple Control Algorithms	6
	1.5	Classical Theory: Optimal and Robust Control	8
	1.6	The Need for a New Theory	9
	1.7	Bibliographic Remarks	12
2	**Dynamical Systems**		13
	2.1	Examples of Dynamical Systems	13
	2.2	Solution Concepts for Dynamical Systems	17
	2.3	Intractability of Equilibrium, Stabilizability, and Controllability	20
	2.4	Bibliographic Remarks	22
	2.5	Exercises	23
3	**Markov Decision Processes**		25
	3.1	Reinforcement Learning	25
	3.2	Markov Decision Processes	28
	3.3	The Bellman Equation	34

	3.4	The Value Iteration Algorithm	36
	3.5	Bibliographic Remarks	38
	3.6	Exercises	39
4	**Linear Dynamical Systems**	40
	4.1	Approximating General Dynamics via LTV Systems	41
	4.2	Stabilizability of Linear Systems	41
	4.3	Controllability of Linear Dynamical Systems	44
	4.4	Quantitative Definitions	46
	4.5	Bibliographic Remarks	48
	4.6	Exercises	48
5	**Optimal Control of Linear Dynamical Systems**	50
	5.1	The Linear Quadratic Regulator	50
	5.2	Optimal Solution of the LQR	51
	5.3	Infinite-Horizon LQR	53
	5.4	Robust Control	55
	5.5	Bibliographic Remarks	56
	5.6	Exercises	56

		Part II Basics of Online Control	57
6	**Regret in Control**	59
	6.1	Online Convex Optimization	60
	6.2	Regret for Control	62
	6.3	Expressivity of Control Policy Classes	64
	6.4	Bibliographic Remarks	69
	6.5	Exercises	70
7	**Online Nonstochastic Control**	71
	7.1	From Optimal and Robust to Online Control	72
	7.2	The Online Nonstochastic Control Problem	73
	7.3	The Gradient Perturbation Controller	75
	7.4	Bibliographic Remarks	79
	7.5	Exercises	80
8	**Online Nonstochastic System Identification**	82
	8.1	The Role of System Identification in Online Control	82
	8.2	A Nonstochastic Approach to System Identification	83
	8.3	Nonstochastic System Identification	84
	8.4	From Indentification to Nonstochastic Control	87
	8.5	Summary and Key Takeaways	87

		Contents	ix
	8.6	Bibliographic Remarks	88
	8.7	Exercises	89

Part III Learning and Filtering   91

9 Learning in Unknown Linear Dynamical Systems   93
  9.1 Learning in Dynamical Systems   93
  9.2 Online Learning of Dynamical Systems   94
  9.3 Classes of Prediction Rules   95
  9.4 Efficient Learning of Linear Predictors   98
  9.5 Summary   99
  9.6 Bibliographic Remarks   100
  9.7 Exercises   100

10 Kalman Filtering   102
  10.1 Observable Systems   103
  10.2 Optimality of the Kalman Filter   104
  10.3 Bayes-Optimality under Gaussian Noise   108
  10.4 Conclusion   109
  10.5 Bibliographic Remarks   110
  10.6 Exercises   111

11 Spectral Filtering   113
  11.1 Spectral Filtering in One Dimension   114
  11.2 Spectral Predictors   116
  11.3 Online Spectral Filtering   120
  11.4 Conclusion   121
  11.5 Bibliographic Remarks   122
  11.6 Exercises   123

Part IV Online Control with Partial Observation   125

12 Policy Classes for Partially Observed Systems   127
  12.1 Linear Observation Action Controllers   128
  12.2 Linear Dynamic Controllers   130
  12.3 Disturbance Response Controllers   131
  12.4 Summary   134
  12.5 Bibliographic Remarks   135
  12.6 Exercises   136

13 Online Nonstochastic Control with Partial Observation   138
    13.1 The Gradient Response Controller   139
    13.2 Extension to Time-Varying Systems   142
    13.3 Conclusion   143
    13.4 Bibliographic Remarks   143
    13.5 Exercises   144

References   147
Index   155

Preface

This text presents an introduction to an emerging paradigm in the control of dynamical systems and differentiable reinforcement learning called *online nonstochastic control*. The new approach applies techniques from online convex optimization and convex relaxations to obtain new methods with provable guarantees for classical settings in optimal and robust control.

The primary distinction between online nonstochastic control and other frameworks is the objective. In optimal control, robust control, and other control methodologies that assume stochastic noise, the goal is to perform comparably to an offline optimal strategy. In online nonstochastic control, both the cost functions and the perturbations from the assumed dynamical model are chosen by an adversary. Thus, the optimal policy is not defined a priori. Rather, the goal is to attain low regret against the best policy in hindsight from a benchmark class of policies.

This objective suggests the use of the decision-making framework of online convex optimization as an algorithmic methodology. The resulting methods are based on iterative mathematical optimization algorithms and are accompanied by finite-time regret and computational complexity guarantees.

Acknowledgments

This book was a journey that started in Princeton at the Google DeepMind Lab in the summer of 2018. At that time Sham Kakade visited the group, which consisted of the authors together with Naman Agarwal, Brian Bullins, Xinyi Chen, Cyril Zhang, and Yi Zhang. This group worked on various aspects of optimization, reinforcement learning, and dynamical systems. The first paper, in which online nonstochastic control was basically invented, started in room 401 in the computer science building at Princeton University. We remember and cherish the exact moment along with the mathematical writings in chalk on that blackboard.

Ever since then, the gem of an idea turned into a complete theory that is now mainstream in machine learning and control, with numerous applications and refinements. Notable events that contributed to this theory include the following. Max Simchowitz, who was a graduate student then, visited Princeton during 2019–2020, where he collaborated with the group, and the theory was extended to partial observed systems. Paula Gradu, who was then an undergraduate, and Edgar Minasyan, while a graduate student, contributed to the study of time-varying and nonlinear dynamical systems. Anirudha Majumdar joined for several important projects, notably for iterative linear control. Udaya Ghai and Arushi Gupta studied extending the GPC to model-free RL.

Most recently, this theory has been used to design new architectures for sequence prediction. The theory of spectral filtering was extended by the aforementioned members of the group, and newer folks joined, including Daniel Suo and Annie Marsden.

This book was further refined through teaching. Elad Hazan gratefully acknowledges the numerous contributions and insights of the students of the course *computational control theory* delivered at Princeton during COVID over Zoom during 2020–2021. The text was further enhanced by students of *theoretical machine learning* taught at Princeton University in the fall of 2024–2025.

The following colleagues and students were particularly helpful, making insightful comments, finding various typos, and suggesting improvements: Gon Buzaglo, Anand Brahmbhatt, Sofiia Druchyna, and Jonathan Pillow (and many more).

Throughout the writing of this book, we engaged in extensive discussions with ChatGPT, an AI assistant. Although it does not claim authorship, its ability to refine mathematical reasoning and clarify complex topics has been a remarkable collaboration in human–AI interaction. In fact, it wrote this acknowledgment all by itself!

Symbols

General

$\stackrel{\text{def}}{=}$	definition
arg min{}	the argument minimizing the expression in braces
$[n]$	the set of integers $\{1, 2, \ldots, n\}$
$\mathbf{1}_A$	the indicator function, equals one if A is true, else zero

Linear Algebra, Geometry, and Calculus

$\mathbb{R}^d$	d-dimensional Euclidean space
Δ_d	d-dimensional simplex, $\{\sum_i \mathbf{x}_i = 1, \mathbf{x}_i \geq 0\}$
$\mathbb{S}_d$	d-dimensional sphere, $\{\|\mathbf{x}\| = 1\}$
$\mathbb{B}_d$	d-dimensional ball, $\{\|\mathbf{x}\| \leq 1\}$
$\mathbb{R}$	real numbers
$\mathbb{C}$	complex numbers
$\|A\|$	determinant of matrix A
$\mathbf{Tr}(A)$	trace of matrix A
$A \succeq 0$	A is positive semi-definite, $\forall v$, $v^\top A v \geq 0$
$A \succeq B$	$A - B$ is positive semi-definite

Control

f	dynamics function
$\mathbf{x}_t \in \mathbb{R}^{d_\mathbf{x}}$	state at time t
$\mathbf{u}_t \in \mathbb{R}^{d_\mathbf{u}}$	control at time t
$\mathbf{w}_t \in \mathbb{R}^{d_\mathbf{w}}$	perturbation at time t
$\mathbf{y}_t \in \mathbb{R}^{d_\mathbf{y}}$	observation at time t
A_t, B_t, C_t	system matrices for linear dynamical system
γ	BIBO stabilizability $\|\mathbf{w}_t\| \leq 1 \longrightarrow \|\mathbf{x}_t\| \leq \gamma$
h	Number of parameters in a policy class
κ	norm bound on policy parameters
K_t	stabilizing linear controller at time t
$K_r(A, B)$	Kalman controllability matrix of order r
$\hat{K}(A, C)$	Kalman observability matrix

Reinforcement Learning

S, A, γ, R, P	Markov decision process with states S, actions A, discount factor γ, rewards R, and transition probability matrix P
R_t	reward at time t
$\mathbf{P}^a_{ss'}$	probability of transition from s to s' given action a
$\mathbf{v}: S \to \mathbb{R}$	value function
$Q: S \times A \to \mathbb{R}$	Q function
$\pi: S \to A$	policy

Optimization

$\mathbf{x}$	vectors in the decision set
$\mathcal{K}$	decision set
$\nabla^k f$	the kth differential of f; note $\nabla^k f \in \mathbb{R}^{d^k}$
$\nabla^{-2} f$	the inverse Hessian of f
∇f	the gradient of f
∇_t	the gradient of f at point $\mathbf{x}_t$
$\mathbf{x}^\star$	the global or local optima of objective f
h_t	objective value distance to optimality, $h_t = f(\mathbf{x}_t) - f(\mathbf{x}^\star)$
d_t	Euclidean distance to optimality $d_t = \|\mathbf{x}_t - \mathbf{x}^\star\|$
G	upper bound on norm of subgradients
D	upper bound on Euclidean diameter
D_p, G_p	upper bound on the p-norm of the subgradients/diameter

PART I

BACKGROUND IN CONTROL AND REINFORCEMENT LEARNING

1
Introduction

1.1 What Is This Book About?

Control theory is the engineering science of manipulating physical systems to achieve the desired functionality. Its history is centuries old and spans rich areas from the mathematical sciences. A brief list includes differential equations, operator theory and linear algebra, functional analysis, harmonic analysis, and more. However, it mostly does not concern modern algorithmic theory or computational complexity from computer science.

The advance of the deep learning revolution has brought to light the power and robustness of algorithms based on gradient methods. The following quote is attributed to deep learning pioneer Yann Lecun during a Barbados workshop on deep learning theory circa February 2020.

"Control = reinforcement learning with gradients."

This book concerns this exact observation: *How can we exploit the differentiable structure in natural physical environments to create efficient and robust gradient-based algorithms for control?*

The answer we give is **not** to apply gradient descent to any existing algorithm. Rather, we rethink the paradigm of control from the start. What can be achieved in a given dynamical system, whether known or unknown, fully observed or not, linear or nonlinear, and without assumptions on the noise sequence?

This is a deep and difficult question that we fall short of answering. However, we do propose a different way of answering this question: We define the control problem from the point of view of an online player in a repeated game. This gives rise to a simple but powerful framework called nonstochastic control theory. The framework is a natural fit to applying techniques from online convex optimization, resulting in new, efficient, gradient-based control methods.

1.2 The Origins of Control

The industrial revolution was in many ways shaped by the steam engine that came to prominence in the eighteenth century. One of its most critical components was introduced by the Scottish inventor James Watt: the *centrifugal governor*, which is amongst the earliest examples of a control mechanism. A major source of efficiency losses of the earlier designs was the difficulty in regulating the engine. Too much or too little steam could easily cause engine failure. The centrifugal governor was created to solve this problem (Figure 1.1).

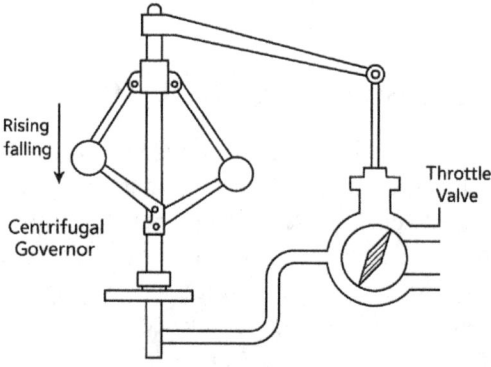

Figure 1.1 A centrifugal governor.

The centrifugal governor consists of a shaft connected to two metal balls that are allowed to rotate. The increased shaft speed causes the balls to spin faster, and the resulting centrifugal force causes the balls to lift. The balls are in turn attached to a pin that controls the amount of steam entering the engine, so as the balls spin faster, the pin lifts and limits the amount of fuel that can enter the system. However, if the balls spin slower, the pin is lowered, allowing more steam to enter the system. This provides natural, direct, and proportional control to the system.

The impact of the centrifugal governor on the steam engine, and in turn on the industrial revolution, illustrates the importance of the field of control in engineering and the sciences. Since the eighteenth century, our world has changed quite a bit, and the importance of control has only grown.

The challenges of controlling systems have moved from the steam engine to the design of aircraft, robotics, and autonomous vehicles. The importance of designing efficient and robust methods remains as crucial as in the industrial revolution.

1.3 Formalization and Examples of a Control Problem

With the physical example of the centrifugal governor in mind, we give a mathematical definition of a control problem. The first step is to formally define a dynamical system in the following way.

Definition 1.1 (Dynamical system) A dynamical system evolving from an initial state $\mathbf{x}_1$ is given by the following equations:

$$\mathbf{x}_{t+1} = f_t(\mathbf{x}_t, \mathbf{u}_t, \mathbf{w}_t), \quad \mathbf{y}_t = g_t(\mathbf{x}_t).$$

Here $\mathbf{x}_t \in \mathbb{R}^{d_x}$ is the state of the system, $\mathbf{y}_t \in \mathbb{R}^{d_y}$ is the observation, $\mathbf{u}_t \in \mathbb{R}^{d_u}$ is the control input, and $\mathbf{w}_t \in \mathbb{R}^{d_w}$ is the perturbation.

The function g_t is a mapping from the state to the observed state, and the function f_t is the transition function, given the current state $\mathbf{x}_t$, the input control $\mathbf{u}_t$, and the perturbation $\mathbf{w}_t$. The subscript indicates the relevant quantities at time step t.

Given a dynamical system, there are several possible objectives that we can formulate. We give a generic formulation below that captures many common control formulations, notably the optimal control problem and the robust control problem. These will be addressed in much more detail later in this book. In this problem formulation, the goal is to minimize the long-term horizon cost given by a sequence of cost functions.

Definition 1.2 (A generic control problem) Given a dynamical system with starting state $(\mathbf{x}_1, \mathbf{y}_1)$, iteratively construct controls u_t to minimize a sequence of loss functions $c_t(\mathbf{y}_t, \mathbf{u}_t)$, $c_t : \mathbb{R}^{d_y + d_u} \to \mathbb{R}$. This is given by the following mathematical program:

$$\min_{\mathbf{u}_{1:T}} \sum_t c_t(\mathbf{y}_t, \mathbf{u}_t)$$

$$\text{s.t. } \mathbf{x}_{t+1} = f_t(\mathbf{x}_t, \mathbf{u}_t, \mathbf{w}_t), \quad \mathbf{y}_t = g_t(\mathbf{x}_t).$$

At this point, many aspects of the generic problem have not been specified. These include:

- Are the transition functions and observation functions known or unknown to the controller?
- How accurately is the transition function known?
- What is the mechanism that generates the perturbations $\mathbf{w}_t$?
- Are the cost functions known or unknown ahead of time?

Different theories of control differ by the answers they give to the above questions, and these in turn give rise to different methods.

Instead of postulating a hypothesis about how our world should behave, we proceed to give an intuitive example of a control problem. We can then reason about the different questions and motivate the nonstochastic control setting.

1.3.1 Example: Control of a Medical Ventilator

A recent relevant example of control is in the case of medical ventilators. The ventilator connects the trachea in the lungs to a central pressure chamber, which helps maintain positive-end expiratory pressure (PEEP) to prevent the lungs from collapsing and simulate healthy breathing in patients with respiratory problems.

The ventilator must take into account the structure of the lung to determine the optimal pressure to induce. Such structural factors include *compliance*, or the change in lung volume per unit pressure, and *resistance*, or the change in pressure per unit flow.

A simplistic formalization of the dynamical system of the ventilator and lung can be derived from the physics of a connected two-balloon system. The dynamics equations can be written as

$$p_t = C_0 + C_1 v_t^{-1/3} + C_2 v_t^{5/3},$$
$$v_{t+1} = v_t + \Delta_t u_t,$$

where p_t is the observed pressure, v_t is the volume of the lung, u_t is the input flow (control), and Δ_t is the unit of time resolution of the device. C_0, C_1, and C_2 are constants that depend on the particular system/lung–ventilator pair.

The goal of ventilator control is to regulate the pressure according to some predetermined pattern. As a dynamical system, we can write $v_{t+1} = f(v_t, u_t, w_t)$, $p_t = g(v_t)$, and we can define the cost function to be $c_t(p_t, u_t) = |p_t - p_t^*|^2$, where p_t^* is the desired pressure at time t. The goal, then, is to choose a sequence of control signals $\{u_t\}_{t=1}^T$ to minimize the total cost over T iterations.

In Figure 1.2 we see an illustration of a simple controller, which we will soon describe, called the PID controller, as it performs on a ventilator and lung simulator.

1.4 Simple Control Algorithms

The previous section detailed the importance of control problems, specifically through the example of a critical care medical device. In this section, we detail

1.4 Simple Control Algorithms

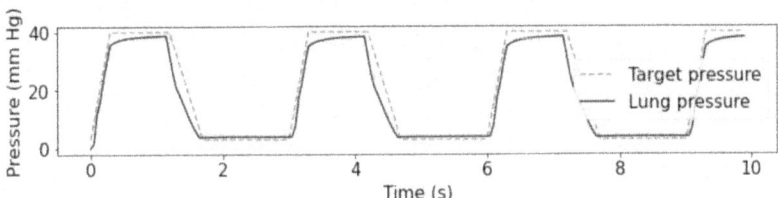

Figure 1.2 Performance of the PID controller on a mechanical ventilator, from Suo et al. (2021).

methods for controlling a medical ventilator, or indeed any physical device that can be modeled as a dynamical system.

The following methods are simplistic but still widely used. For example, the PID algorithm is still used in most medical ventilators today. In later sections, we describe a modern optimization-based framework that yields more sophisticated algorithms that are capable of better performance.

1.4.1 The Bang-Bang Controller

Bang-bang control is one of the simplest control strategies. Consider a scalar dynamical system, where it is desirable to keep the state x_t within a prespecified dynamic range $[x_{\min}, x_{\max}]$ (which can vary over time) and where the scalar control input u_t must at all times be within the interval $[u_{\min}, u_{\max}]$, which is assumed to contain zero for simplicity. A bang-bang controller chooses the control input at time t as

$$u_t = \mathbf{1}_{\{x_t < x_{\min}\}} \times u_{\max} + \mathbf{1}_{\{x_t > x_{\max}\}} \times u_{\min}.$$

A bang-bang controller saturates its inputs to extreme values in an attempt to confine the state of the system to a desirable range. For example, in the context of mechanical ventilation, one could choose $[p_t^* - \Delta p, p_t^* + \Delta p]$ as a reasonable region around the prescribed pressure p_t^* as a desirable range for p_t. In such a case, a bang-bang controller would allow oxygen to enter at the highest possible rate whenever the pressure drops below $p_t^* - \Delta p$.

This discussion highlights two unsatisfactory aspects of the bang-bang control scheme: one, it ignores the cost associated with acting with the extreme values of the control input; two, it produces oscillatory and rapidly changing behavior in the control state.

1.4.2 The PID Controller

A more sophisticated control algorithm is the PID control method. Consider a control method that chooses the control to be a linear function of the observed states, that is:

$$\mathbf{u}_t = \sum_{i=0}^{k} A_i \mathbf{x}_{t-i},$$

for some coefficients A_i, where we consider only the $k+1$ most recent states. One of the most useful special cases of this linear family is the PID controller.

A special subfamily of linear controllers is that of linear controllers that optimize on a predefined basis of historical observations. More precisely,

(i) Proportional control (e.g., centrifugal governor): $\mathbf{u}_t = \alpha_0 \mathbf{x}_t$.
(ii) Integral control: $\mathbf{u}_t = \beta \sum_{i<t} \mathbf{x}_{t-i}$.
(iii) Derivative control: $\mathbf{u}_t = \gamma(\mathbf{x}_t - \mathbf{x}_{t-1})$.

The PID controller, named after the nature of the three components that make up it, specifies three coefficients α, β, γ to generate a control signal. In that sense, it is extremely sparse, which from a learning-theoretic perspective is a good indication of generalization.

In fact, the PID controller is the method of choice for numerous engineering applications, including control of medical ventilators.

1.5 Classical Theory: Optimal and Robust Control

Classical theories of control differ within the generic control problem formulation 1.2 primarily in one significant aspect: the perturbation model. A rough divide between the two main theories – *optimal control* and *robust control* – is as follows.

Optimal control postulates a probabilistic model for the perturbations, which are typically independent and identically distributed (i.i.d.) draws from a Gaussian distribution. However, other types of distribution can also be considered. As such, the control problem can be rephrased in a more precise stochastic mathematical program,

$$\min_{\mathbf{u}_{1:T}} \mathbb{E}_{\mathbf{w}_{1:T}} \left[\sum_t c_t(\mathbf{y}_t, \mathbf{u}_t) \right]$$
$$\text{s.t.} \ \mathbf{x}_{t+1} = f_t(\mathbf{x}_t, \mathbf{u}_t, \mathbf{w}_t), \quad \mathbf{y}_t = g(\mathbf{x}_t).$$

In contrast, robust control theory plans for the worst-case noise from a given set of constraints, denoted $\mathcal{K}$. Thus, the robust control problem can be rephrased as the following mathematical program:

$$\min_{\mathbf{u}_{1:T}} \max_{\mathbf{w}_{1:T} \in \mathcal{K}} \left[\sum_t c_t(\mathbf{y}_t, \mathbf{u}_t) \right]$$
$$s.t. \quad \mathbf{x}_{t+1} = f_t(\mathbf{x}_t, \mathbf{u}_t, \mathbf{w}_t), \quad \mathbf{y}_t = g(\mathbf{x}_t).$$

The two formulations have been debated for decades, with pros and cons for each side. Optimal control focuses on the average case and often yields simpler formulations for optimization, resulting in more efficient methods. Robust control, in contrast, allows for more difficult noise models and some amount of model misspecification. However, it is too pessimistic for many natural problems.

Nonstochastic control theory takes the best of both worlds, simultaneously granting robustness to adversarial noise and allowing the use of optimistic behavior when appropriate while providing computationally efficient methods.

1.6 The Need for a New Theory

Consider the problem of flying a drone from the source to the destination, subject to unknown weather conditions. The aerodynamics of flight can be modeled sufficiently well by time-varying linear dynamical systems, and existing techniques are perfectly capable of doing a great job for indoor flight. However, the wind conditions, rain, and other uncertainties are a different story. Certainly, the wind is not an i.i.d. Gaussian random variable! Optimal control theory, which assumes this zero-mean noise, is therefore overly optimistic and inaccurate for our problem.

The designer might resort to robust control theory to account for all possible wind conditions, but this is overly pessimistic. What if we encounter a bright sunny day after all? Planning for the worst case would mean slow and fuel-inefficient flight.

We would like an adaptive control theory that allows us to attain the best of both worlds: an efficient and fast flight when the weather permits, and careful drone maneuvering when this is required by the conditions. Can we design a control theory that will allow us to take into account the specific instance perturbations and misspecifications, and give us finite-time provable guarantees? This is the subject of online nonstochastic control!

1.6.1 Online Nonstochastic Control Theory

This book is concerned with nonstochastic noise. When dealing with nonstochastic, that is, arbitrary or even adversarial, perturbations, the optimal policy, that is, a decision-making rule mapping observations to control inputs, is not clear a priori. Rather, an optimal policy for the observed perturbations can be determined *only in hindsight*. This is a significant difference from optimal and robust control theory, where the mapping from states to actions can be completely predetermined based upon the noise model.

Since optimality is defined only in hindsight, the goal is not predetermined. We thus shift to a different performance metric borrowed from the world of learning in repeated games, namely regret. Specifically, we consider regret with respect to a reference class of policies. In online control, the controller iteratively chooses a control $\mathbf{u}_t$. The controller then observes the next state of the system $\mathbf{x}_{t+1}$ and suffers a loss of $c_t(\mathbf{x}_t, \mathbf{u}_t)$ according to an adversarially chosen loss function. For simplicity, we consider the case of full state observation where $\mathbf{y}_t = \mathbf{x}_t$. Let $\Pi = \{\pi : \mathbf{x} \to \mathbf{u}\}$ be a class of policies. The regret of the controller with respect to Π is defined as follows.

Definition 1.3 Given a dynamical system, the regret of an online control algorithm $\mathcal{A}$ over T iterations with respect to a class of policies Π is given by

$$\mathrm{regret}_T(\mathcal{A}, \Pi) = \max_{\mathbf{w}_{1:T} : \|\mathbf{w}_t\| \le 1} \left(\sum_{t=1}^T c_t(\mathbf{x}_t, \mathbf{u}_t) - \min_{\pi \in \Pi} \sum_{t=1}^T c_t(\mathbf{x}_t^\pi, \mathbf{u}_t^\pi) \right),$$

where $\mathbf{u}_t = \mathcal{A}(\mathbf{x}_t)$ are the controls generated by $\mathcal{A}$, and $\mathbf{x}_t^\pi, \mathbf{u}_t^\pi$ are the counterfactual state sequence and controls under the policy π, that is:

$$\mathbf{u}_t^\pi = \pi(\mathbf{x}_t^\pi),$$
$$\mathbf{x}_{t+1}^\pi = f_t(\mathbf{x}_t^\pi, \mathbf{u}_t^\pi, \mathbf{w}_t).$$

Henceforth, if T, Π, and $\mathcal{A}$ are clear from the context, we drop these symbols when defining regret.

At this point, the reader may wonder why we should compare to a reference class, as opposed to all possible policies. There are two main answers to this question:

(i) First, the best policy in hindsight out of all possible policies may be very complicated to describe and to reason about. This argument was made by the mathematician Tyrell Rockafellar; see the bibliographic section for more details.

(ii) Second, it can be shown that it is impossible, in general, to obtain sublinear regret versus the best policy in hindsight. Rather, a different performance

1.6 The Need for a New Theory

metric called *competitive ratio* can be analyzed, as we discuss later in the book.

The nonstochastic control problem can now be stated as finding an efficient algorithm that minimizes the worst-case regret vs. meaningful policy classes that we examine henceforth. It is important to mention that *the algorithm does not have to belong to the comparator set of policies!* Indeed, in the most powerful results, we will see that the algorithm will learn and operate over a policy class that is strictly larger than the comparator class.

1.6.2 A New Family of Algorithms

The policy regret minimization viewpoint in control leads to a new approach to algorithm design. Instead of a priori computation of the optimal controls, online convex optimization suggests modifying the controller according to the costs and dynamics, in the flavor of adaptive control, in order to achieve provable bounds on the policy regret.

As a representative example, consider the *gradient perturbation controller* described in Algorithm 1.1. The algorithm maintains matrices that are used to create the control as a linear function of past perturbations, according to line 3. These parameters change over time according to a gradient-based update rule, where the gradient is calculated as the derivative of a counterfactual cost function with respect to the parametrization, according to line 5.

Algorithm 1.1 Gradient Perturbation Controller (GPC) – Simplified Version
---
1: Input: h, η, initial parameters $M_{1:h}^1$, dynamics f.
2: **for** $t = 1 \ldots T$ **do**
3:   Use control $\mathbf{u}_t = \sum_{i=1}^{h} M_i^t \mathbf{w}_{t-i}$
4:   Observe state $\mathbf{x}_{t+1}$, compute perturbation $\mathbf{w}_t = \mathbf{x}_{t+1} - f(\mathbf{x}_t, \mathbf{u}_t)$
5:   Construct loss $\ell_t(M_{1:h}) = c_t(\mathbf{x}_t(M_{1:h}), \mathbf{u}_t(M_{1:h}))$
     Update $M_{1:h}^{t+1} \leftarrow M_{1:h}^t - \eta \nabla \ell_t(M_{1:h}^t)$
6: **end for**

The version given here assumes that the time-invariant dynamics f are known. The GPC computes a control that is a linear function of past perturbations. It modifies the linear parameters according to the loss function via an iterative gradient method. In Chapter 7, we prove that this algorithm attains sublinear regret bounds for a large class of time-varying and time-invariant dynamical systems. In later chapters, we explore extensions to nonlinear dynamical systems and unknown systems, which lie at the heart of nonstochastic control theory.

1.7 Bibliographic Remarks

Control theory is a mature and active discipline with both deep mathematical foundations and a rich history of applications. In terms of application, self-regulating feedback mechanisms that limit the flow of water, much like the modern float valve, were reportedly used in ancient Greece over two thousand years ago. The first mathematical analysis of such systems appeared in a paper by James Clerk Maxwell on the topic of governors (Maxwell, 1868), shortly after Maxwell had published his treatise on the unification of electricity and magnetism. For an enlightening historical account, see Fernández Cara and Zuazua Iriondo (2003).

For a historical account on the development of the PID controller, see Bennett (1993). The theoretical and practical properties of this controller are described in detail in Åström and Hägglund (1995).

There are many excellent textbooks on control theory, which we point to in later chapters. The book of Bertsekas (2007) offers extensive coverage of topics from the point of view of dynamic programming. The texts Stengel (1994) and Zhou et al. (1996a), in contrast, delve deeply into results on linear control, the most well-understood branch of control theory. Rockafellar (1987) proposed the use of convex cost functions to model the cost in a linear dynamical system to incorporate constraints on state and control and suggested computational difficulties with this approach. The text Slotine and Li (1991) gives a thorough introduction to adaptive control theory.

Online nonstochastic control theory and the gradient perturbation control algorithm were proposed in Agarwal et al. (2019c). A flurry of novel methods and regret bounds for online nonstochastic control have appeared in recent years, many of which are surveyed in Hazan and Singh (2021). Notable among them are logarithmic regret algorithms (Agarwal et al., 2019b; Foster and Simchowitz, 2020; Simchowitz, 2020), nonstochastic control for unknown systems (Hazan et al., 2020), black-box online nonstochastic control (Chen and Hazan, 2021), and nonstochastic control with partial observability (Simchowitz et al., 2020).

The theory is heavily based on recent developments in decision making and online learning, notably the framework of online convex optimization (Hazan, 2016) and especially its extensions to learning with memory (Anava et al., 2015).

2
Dynamical Systems

In this chapter, we return to the general problem of control from Definition 1.2, that is, choosing a sequence of controls to minimize a sequence of cost functions. Before moving on to algorithms for control, it is important to reason about what kind of solution is even possible.

For this reason, we consider several natural examples of dynamical systems and reason about what constitutes desirable behavior. This will help us understand the kinds of observation models and control objectives that are feasible and natural in real applications.

2.1 Examples of Dynamical Systems

2.1.1 Medical Ventilator

In the task of mechanical ventilation, the control algorithm dictates the amount of oxygen inflow u_t (in m^3/s) into the lungs of a patient. This inflow approximately determines the volume of the lungs v_t via a simple linear relation,

$$v_{t+1} = v_t + \Delta u_t,$$

where Δ is the time elapsed between successive actions of the discrete-time control algorithm. Being difficult to directly measure or estimate, the quantity v_t is a hidden or latent variable from the controller's perspective. Instead, the controller observes the air pressure p_t. We give a simplified physical model that relates the pressure and the volume. This model, given in the following, can be derived by first principles from stress–strain relations, treating the left and right lungs connected by the trachea as situated in a two-balloon experiment:

$$p_t = C_0 + C_1 v_t^{-1/3} + C_2 v_t^{5/3}.$$

The objective of the control algorithm is to follow an ideal pressure trajectory p_t^* prescribed by the physician as closely as possible, say, as measured by squared loss $c_t(p_t, u_t) = |p_t - p_t^*|^2 + |u_t|^2$, while limiting the maximum instantaneous airflow to minimize patient discomfort. See Figure 2.1 for a schematic representation of airflow when a lung is connected to a mechanical ventilator.

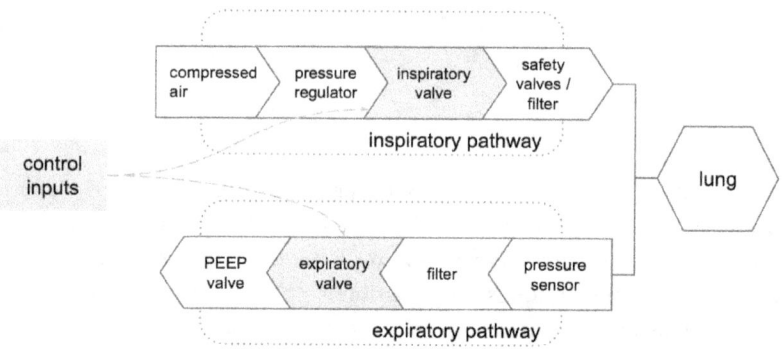

Figure 2.1 A schematic of the respiratory circuit from Suo et al. (2021).

2.1.2 Double Integrator

The double integrator, as depicted in Figure 2.2, is a common example of a *linear* dynamical system. In its simplest form, it models a one-dimensional point object moving according to Newtonian mechanics on a one-dimensional trajectory (i.e. along the real number line). The state $\mathbf{x}_t$ is two-dimensional, with the first coordinate indicating the position of a point object and the second denoting its velocity. The control algorithm decides the force $u_t \in \mathbb{R}$ that is applied to the object at each time step. The dynamics equation can be written as

$$\mathbf{x}_{t+1} = \begin{bmatrix} 1 & \Delta \\ 0 & 1 \end{bmatrix} \mathbf{x}_t + \begin{bmatrix} 0 \\ 1 \end{bmatrix} u_t + \mathbf{w}_t.$$

The position of the object is linearly related to its velocity via the parameter Δ, which measures the time resolution the controller operates in. Optionally, the state is also affected by external perturbations $\mathbf{w}_t$ outside the influence of the controller.

The goal here is to ensure that the point object arrives and then comes to rest at some coordinate $x^* \in \mathbb{R}$ as quickly as possible while also not expending too much energy. The cost function is naturally expressed as

2.1 Examples of Dynamical Systems

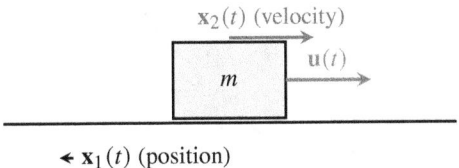

Figure 2.2 Double integrator illustration showing state coordinate $x_1(t)$ (position), state coordinate $x_2(t)$ (velocity), mass m, and control input $u(t)$.

$$c(\mathbf{x}_t, u_t) = \left\| \mathbf{x}_t - \begin{bmatrix} x^* \\ 0 \end{bmatrix} \right\|^2 + u_t^2.$$

2.1.3 Pendulum Swing-Up

The swing-up pendulum (Figure 2.3) is one of the simplest examples of rotational kinematics. It corresponds to a point object of mass m mounted at the end of a massless rod of length l. The other end of the rod is fixed to a pivot. Being vertically aligned, the system is subject to gravity with gravitational acceleration g. The controller is responsible for applying a torque $u_t \in \mathbb{R}$ at each time step. Given the vertical angle θ_t of the rod, the physics of this system may be summarized as

$$\theta_{t+1} = \theta_t + \Delta \dot{\theta}_t, \qquad \dot{\theta}_{t+1} = \dot{\theta}_t + \Delta \frac{u_t - mg \sin \theta_t}{ml^2},$$

where Δ is the time resolution of the controller. Note that this dynamical equation is nonlinear in the state $\mathbf{x}_t = [\theta_t, \dot{\theta}_t]^\top$ due to the presence of the term $\sin \theta$ in the evolution of $\dot{\theta}_t$.

The objective here is to get the rod to rest in a vertical upright position at an angle of π, using as little torque as possible throughout the process. This may be expressed through the cost function

$$c(\mathbf{x}_t, u_t) = \left\| \mathbf{x}_t - \begin{bmatrix} \pi \\ 0 \end{bmatrix} \right\|^2 + u_t^2.$$

2.1.4 Aircraft Dynamics

A final example of a dynamical system we consider is the longitudinal (vertical tilt) motion of an aircraft that is regulated by automated controllers; see schematic description in Figure 2.4. We present the approximate dynamics for the Boeing 747 aircraft, which is a linear dynamical system, from Boyd (2010).

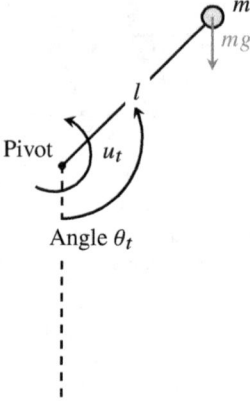

Figure 2.3 Pendulum swing-up illustration with clearly marked angle θ_t, gravitational force mg, control input torque u_t, mass m, and rod length l.

The two-dimensional control input $\mathbf{u}_t$ corresponds to the elevation angle and the thrust. The disturbances correspond to the velocity of the wind along the body axis and perpendicular to it.

The following dynamics equation has matrices whose numerical values were computed using the dynamics equation for an actual aircraft. These numerical values can be determined by the mass, length, and other physical properties of the aircraft:

$$\mathbf{x}_{t+1} = \begin{bmatrix} -0.003 & 0.039 & 0 & -0.322 \\ -0.065 & -0.319 & 7.74 & 0 \\ 0.020 & -0.101 & -0.429 & 0 \\ 0 & 0 & 1 & 0 \end{bmatrix} \mathbf{x}_t + \begin{bmatrix} 0.01 & 1 \\ -0.18 & -0.04 \\ -1.16 & 0.598 \\ 0 & 0 \end{bmatrix} \mathbf{u}_t$$
$$+ \begin{bmatrix} -0.003 & 0.039 & 0 & -0.322 \\ -0.065 & -0.319 & 7.74 & 0 \end{bmatrix} \mathbf{w}_t.$$

The quantities of interest for assessing the performance of the controller are the aircraft speed, climb rate – each of which is a linear transformation of the state – and applied thrust. The corresponding cost is

$$c(\mathbf{x}_t, \mathbf{u}_t) = \left\| \begin{bmatrix} 1 & 0 & 0 & 0 \\ 0 & -1 & 0 & 7.74 \end{bmatrix} \mathbf{x}_t - \mathbf{x}_t^* \right\|^2 + \left\| \begin{bmatrix} 0 & 1 \end{bmatrix} \mathbf{u}_t \right\|^2.$$

2.2 Solution Concepts for Dynamical Systems

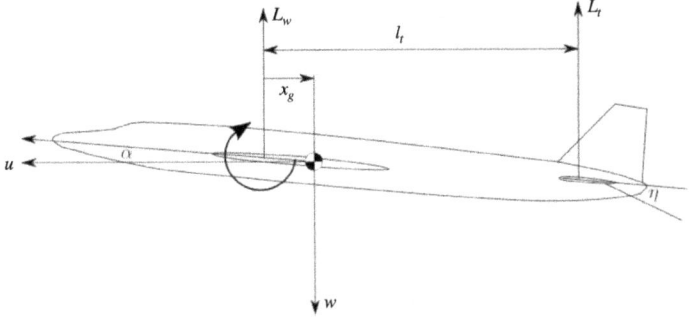

Figure 2.4 Longitudinal control in an aircraft. Graphic by User_A1, licensed under CC-BY-SA 4.0; it is an SVG derivative of a public domain work by G J Vigurs.

2.1.5 Epidemiological Models

The COVID-19 pandemic marked an increase in public interest in policy discussions on the effectiveness of certain interventions as a means of controlling infection spread. Epidemiological models predict the spread of an infectious disease via structural models and differential equations. We present a discrete-time analogue of the most basic of such models, the SIR model. Let S_t, I_t, and R_t denote the population sizes of those susceptible, infected, and recovered, respectively. Let u_t be the number of vaccines administered at time t. The dynamics of the basic SIR model are given by the equations below and depicted in Figure 2.5.

$$S_{t+1} = S_t - \beta S_t I_t - \alpha u_t,$$
$$I_{t+1} = I_t + \beta S_t I_t - \gamma I_t,$$
$$R_{t+1} = R_t + \gamma I_t.$$

Here, the coefficient β measures the level of communicability of the infection, γ measures the rate at which an infected person recovers, and α represents the efficacy of the vaccine in preventing infections.

2.2 Solution Concepts for Dynamical Systems

Before considering the control, or manipulation, of a dynamical system, we must first reason about what the goal should even be. In this section, we discuss various solution concepts for a dynamical system, including equilibrium, stability, stabilizability, and controllability.

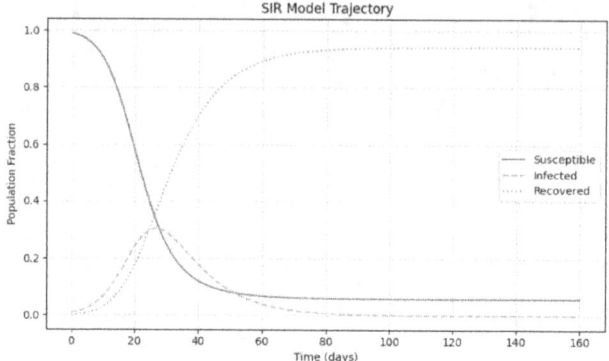

Figure 2.5 Trajectory of a continuous-time SIR model.

Given a dynamical system $x_{t+1} = f(x_t, u_t, w_t)$, the equivalent deterministic dynamical system, defined through the dynamics $x_{t+1} = f'(x_t, u_t) = f(x_t, u_t, 0)$, models the evolution of the dynamical system in the absence of perturbations. In addition, the autonomous dynamical system $x_{t+1} = f''(x_t) = f(x_t, 0, 0)$ describes the state transitions of the dynamical system in the absence of both control inputs and perturbations.

2.2.1 Equilibrium of Dynamical Systems

An example of an equilibrium is a ball sitting at rest at the bottom of a pit. If the ball were to be slightly moved, it would return to the bottom of the pit. Thus, this equilibrium is stable and attractive.

If the ball is at rest on the top of a mountain, pushing the ball in any direction would cause it to fall. This is an equilibrium that is unstable. A formal definition is as follows.

Definition 2.1 A point x is an equilibrium point of an autonomous dynamical system f if and only if $x = f(x)$. An equilibrium point x is stable and attractive if and only if $x = \lim_{t \to \infty} f^t(x_0)$ for every initial state x_0 in some neighborhood of x. If the neighborhood includes the entire set of states, we say that the equilibrium is globally attractive.

Another example is given by a linear transformation in Euclidean space. Consider the system

$$x_{t+1} = Ax_t.$$

Clearly 0 is an equilibrium point as $A0 = 0$ for all A. If the modulus of all eigenvalues is strictly less than one, then after sufficiently many applications of A, any vector converges to the null vector. Thus, $\mathbf{0}$ is a **globally attractive and stable equilibrium**.

In general, determining whether a system has an equilibrium is an intractable problem, as we show in this chapter.

2.2.2 Stabilizability and Controllability

Two fundamental properties of dynamical systems are stabilizability and controllability, as defined next.

Definition 2.2 A dynamical system is **stabilizable** if and only if, for every initial state, there exists a sequence of controls that drives the deterministic-equivalent system to the zero state. That is, $\lim_{t \to \infty} \mathbf{x}_t = \mathbf{0}$.

Definition 2.3 A dynamical system is **controllable** if and only if, for every initial and target state, there exists a sequence of controls that drives the equivalent deterministic system to the target state.

Note that controllability clearly implies stabilizability, as the zero state is one possible target state.

In the next section, we discuss computational difficulties that arise in determining if a system is stabilizable and/or controllable.

2.2.3 Quantitative Definitions of Stabilizability

The quantified notion of stabilizability we introduce below is more refined than driving the system to zero. We require the ability to drive the system to zero under **adversarial** bounded noise in the dynamics. In classical control theory this is called **BIBO-stability**, which stands for bounded-input-bounded-output.

This definition can apply to any dynamical system and is an important concept in nonstochastic control.

Definition 2.4 (BIBO-stability) A control policy π for a dynamical system f is γ-**stabilizing** if and only if the following holds. For every starting state $\|\mathbf{x}_0\| \leq 1$ and every sequence of adversarially chosen bounded perturbations $\mathbf{w}_{1:T}$ such that $\|\mathbf{w}_t\| \leq 1$ for all $t \in [T]$, let $\mathbf{x}_t^\pi$ be the state obtained by using policy π,

$$\mathbf{x}_{t+1}^\pi = f(\mathbf{x}_t^\pi, \pi(\mathbf{x}_t), \mathbf{w}_t).$$

Then the norm of the system's state at every iteration is bounded by an absolute constant independent of T, that is,

$$\forall t, \ \|\mathbf{x}_t^\pi\| \leq \gamma.$$

A dynamical system is said to be γ-stabilizable if and only if it admits a γ-stabilizing control policy.

2.3 Intractability of Equilibrium, Stabilizability, and Controllability

Before formally proving the computational hardness of determining various solution concepts of dynamical systems, we start with a few intuitive and convincing examples of the complexities that arise. Consider the following dynamical system over the complex numbers:

$$z_{t+1} = f_c(z_t) = z_t^2 + c.$$

The Julia set is now defined as

$$J_c = \{z \in C : \limsup_{t \to \infty} |f_c^t(z)| < \infty\},$$

where $f_c^t(x) = f_c(f_c(\ldots f_c(x)))$ denotes t successive applications of the function f_c. In other words, it is the set of all complex numbers Z for which the system has a stable attractive equilibrium (that is not ∞). The boundary of the Julia set is very hard to determine, and it is known that determining if a ball around a given point belongs to the Julia set is a computationally undecidable problem, that is, there is no Turing machine that can decide it in finite time! (See, e.g., Braverman and Yampolsky (2008)). The complexity that arises near the boundary of the Julia set is depicted in Figure 2.6.

Another compelling visual illustration of the complexity of equilibrium in dynamical systems is the Lorenz attractor. This is a strange equilibrium resulting from simple dynamics that can be described using the following three equations over three variables, called the Lorenz equations (Lorenz, 1963),

$$x_{t+1} = \sigma(y_t - x_t),$$
$$y_{t+1} = x_t(\rho - z_t) - y_t,$$
$$z_{t+1} = x_t y_t - \beta z_t.$$

The three-dimensional (3D) trajectories of this system are attracted to a 2D manifold with two "wings" that resemble a butterfly, as depicted in Figure 2.7. For formal properties of this attractor set, see Strogatz (2014) and Tucker (1999). This shows that even studying a solution concept itself is a formidable task.

For a formal proof of attractiveness in the continuous case, see section 9.2 of Strogatz (2014) or the original paper of Lorenz (1963).

2.3 Intractability of Equilibrium, Stabilizability, and Controllability

Figure 2.6 Julia set.

Figure 2.7 Lorenz attractor. Visualization by D Schwen, licensed under CC-BY-2.5.

2.3.1 Computational Hardness of Stabilizability

Consider the following problem: Given a dynamical system, can we determine if it stabilizable, that is, for every possible starting state, does there exist a choice of control inputs that drives the system to the null state? We now show that this decision problem is computationally hard.

Theorem 2.5 *For a family of dynamical systems described as polynomials with integer coefficients, determining the stabilizability of any member is NP-hard.*

We show that this problem can be reduced to the 3SAT problem, a well-known NP-hard problem. This implies that determining the stabilizability of a dynamical system is NP-hard in general.

Proof We first claim that checking if a given polynomial has a zero is NP-hard. This is achieved by a reduction from 3SAT. Recall the 3SAT problem: Given a Boolean formula in the conjunctive normal form over n variables and with m clauses where each clause is restricted to three literals, such as

$$F(x_{1:n}) = (x_1 \vee \neg x_2 \vee x_3) \wedge (\neg x_1 \vee \neg x_2 \vee x_3) \wedge (x_1 \vee x_3 \vee x_4),$$

decide if there exists a satisfying Boolean assignment.

Given such a formula F, consider the following polynomial $p_F(\mathbf{u})$. For each clause c in F, construct a degree-three polynomial as

$$g_{x_i \vee x_j \vee x_k}(\mathbf{u}) \rightarrow (1-u_1) \cdot (1-u_2) \cdot (1-u_3),$$
$$g_{\neg x_i \vee x_j \vee x_k}(\mathbf{u}) \rightarrow u_1 \cdot (1-u_2) \cdot (1-u_3),$$
$$g_{\neg x_i \vee \neg x_j \vee x_k}(\mathbf{u}) \rightarrow u_1 \cdot u_2 \cdot (1-u_3),$$
$$g_{\neg x_i \vee \neg x_j \vee \neg x_k}(\mathbf{u}) \rightarrow u_1 \cdot u_2 \cdot u_3,$$

and the degree-six polynomial for the entire formula as

$$p_F(\mathbf{u}) = \sum_{i=1}^{n}(u_i(1-u_i))^2 + \sum_{j=1}^{m}(g_{c_j}(\mathbf{u}))^2.$$

For a disjunctive clause to be false, all its literals must evaluate to false. Therefore, given any satisfying Boolean assignment $x_{1:n}$ for the formula F, it holds that $p_F(x_{1:n}) = 0$.

Similarly, given any $\mathbf{u} \in \mathbb{R}^n$ such that $p_F(\mathbf{u}) = 0$, it must be that $\mathbf{u}_i \in \{0, 1\}$ for all i. Furthermore, since p_F is a sum of squares, the Boolean assignment $x_{1:n} = \mathbf{u}$ must satisfy every clause in F. This shows the hardness of determining the existence of a zero for polynomials.

Now, consider the scalar-state dynamical equation given by

$$x_{t+1} = (1-\varepsilon)x_t + p_F(\mathbf{u}_t),$$

with p being the polynomial family we just constructed and $\varepsilon > 0$ an arbitrarily small constant. For the system to reach the null state from an arbitrary start state, we would need to verify if there is an assignment of $\mathbf{u}$ such that $p_F(\mathbf{u}) = 0$, which as we just proved is NP-hard. □

The proof establishes that even for a dynamical system that is a polynomial of degree six, checking for stabilizability is NP-hard. Exercise 2.1 asks the reader to strengthen this result and show that checking the stabilizability of dynamical systems describable as polynomials of degree four is also NP-hard.

2.4 Bibliographic Remarks

The text Tedrake (2020) offers several examples and characterizations of dynamical systems useful for robotic manipulation. A more mathematically involved introduction appears in Strogatz (2014).

Adam (2020) discusses further aspects of simulations used in the early days of the COVID-19 pandemic. See Suo et al. (2021) for examples of control algorithms used for noninvasive mechanical ventilation.

The Lorenz equations were introduced to model the phenomenon of atmospheric convection (Lorenz, 1963). For detailed information on the mathematical properties of the Lorenz attractor, see Strogatz (2014) and Tucker (1999). For a rigorous study of Julia sets and the Mandelbrot set, see Braverman and Yampolsky (2008).

Nonlinear dynamical systems admit a variety of stability criteria; we chose to define a few that are useful for further developments in the text. A notable omission from our discussion is that of local and global stability; a celebrated result (Lyapunov, 1992) proves that the existence of certain *potential functions* is sufficient to guarantee global stability, thus obviating the need for a trajectory-based analysis. An interested reader might refer to Khalil (2015) for an expanded discussion on how these stability criteria compare.

An excellent survey on the computational complexity of control is Blondel and Tsitsiklis (2000). For a more recent and stronger result on stabilizability, see Ahmadi (2012). The NP-hardness theorem in this chapter appeared in Ahmadi (2016). The computational complexity of mathematical programming – both constrained and unconstrained – is commented upon in Hazan (2019).

2.5 Exercises

2.1. Strengthen Theorem 2.5 to show that checking whether a dynamical system that is a polynomial of degree four is stabilizable is NP-hard. Hint: Reduce from the NP-hard problem of MAX-2-SAT instead of 3-SAT.

2.2. Let $A \in \mathcal{R}^{d \times d}$ be a diagonalizable matrix. Consider the following dynamical system:

$$\mathbf{x}_{t+1} = \frac{A \mathbf{x}_t}{\|\mathbf{x}_t\|}.$$

What are the stable and attractive equilibrium points $\mathbf{x} \in \mathcal{R}^d$ of the system? Are they also globally attractive? Your answer may depend on A.

2.3. Let $\alpha, \beta, \gamma > 0$. Does the SIR-model, as described in this chapter, define a stabilizable dynamical system (for an appropriate and fixed definition of the "zero state")? What about controllability? We assume that only starting states with the property $S_0 + I_0 + R_0 = 1$ are allowed and that $S_t, I_t, R_t, u_t \geq 0$ must hold at all times.

2.4. Consider the double integrator system:

$$\mathbf{x}_{t+1} = A\mathbf{x}_t + B u_t + \mathbf{w}_t,$$

where

$$A = \begin{bmatrix} 1 & \Delta \\ 0 & 1 \end{bmatrix}, \quad B = \begin{bmatrix} 0 \\ 1 \end{bmatrix}, \quad \mathbf{x}_t = \begin{bmatrix} x_t \\ v_t \end{bmatrix} \in \mathbb{R}^2, \quad u_t \in \mathbb{R}, \quad \Delta > 0.$$

The goal is to ensure that the system reaches the desired state $[x^*, 0]^\top$ as $t \to \infty$ from any initial condition.

(i) Prove that (A, B) is controllable. That is, show that for any initial state and any target state, there exists a finite sequence of inputs that transitions the system from the initial state to the target state.

(ii) Design a stabilizing state-feedback law
$$u_t = -K(\mathbf{x}_t - [x^*\ 0]^\top),$$
with $K = [k_1\ k_2]$, which places all eigenvalues of $A - BK$ strictly within the unit circle. Find conditions on Δ, k_1, k_2 that ensure that all eigenvalues of $A - BK$ are real and lie strictly between -1 and 1, thus guaranteeing $\mathbf{x}_t \to [x^*, 0]^\top$.

3
Markov Decision Processes

The previous chapters described and motivated the problem of control in dynamical systems. We have seen that in full generality, even simple questions about dynamical systems such as stabilizability and controllability are computationally hard. In this chapter, we consider a different representation of the control problem for which computationally efficient methods exist.

This chapter introduces a setting of equal generality as that of the control problem from the previous chapter: reinforcement learning in Markov decision processes (MDPs). However, the representation of states and dynamics in an MDP is more explicit and verbose. In such a case, we can design efficient methods aimed at optimality given the explicit state and transition representations.

After introducing the reinforcement learning model and Markov decision processes, we discuss an important algorithmic technique that will be useful throughout this book: dynamic programming and the accompanying characterization of optimal policies via the Bellman equation.

3.1 Reinforcement Learning

Reinforcement learning is a subfield of machine learning that formally models the setting of learning through interaction in a reactive environment. This is in contrast to learning from examples, as in supervised learning, or inference according to a probabilistic model, as in Bayesian modeling. In reinforcement learning, we have an agent and an environment. The agent observes its position (or state) in the environment and takes actions that transition it to a new state. The environment looks at the agent's state and hands out rewards based on a set of criteria hidden from the agent.

Typically, the goal of reinforcement learning is for the agent to learn behavior that maximizes the total reward it receives from the environment (see Figure 3.1).

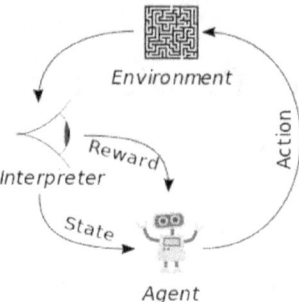

Figure 3.1 The agent–environment interaction loop in reinforcement learning.

This methodology has led to some notable successes: Machines have learned how to play Atari games and how to beat the highest-rated human players at Go.

A crucial assumption for the reinforcement learning model is as follows:

> **The Markovian assumption:** Future states, observations, and rewards are independent of the past given the current state.

The reader can imagine situations where such an assumption is reasonable. For example, in the inverted pendulum setting described in Section 2.1.3, the assumption is clearly valid. It is not significant how a certain state, describing the angle and the angular velocity, was reached for the control problem of how to optimally control the pendulum from this state onward.

Other problems may call into question this assumption. For example, in the medical ventilation problem, it may very well be important how a certain state was reached, as future treatment may depend on the medical history of the patient.

It can be argued that the Markovian assumption can always be made to work by adding more features to the state. In extreme cases, the state can contain the entire history, which makes the assumption true by default. However, this makes the state representation inefficient.

3.1.1 A Classical Example: Bubbo's World

Before diving into the formalism of reinforcement learning, we start with a classical example from the text of Russel and Norvig (2002) (see Figure 3.2). In this example, a robot starts in the bottom left square, and its goal is to

3.1 Reinforcement Learning

reach the +1 reward. The blocks +1, −1 represent terminal states, which means that the game ends when the robot reaches them. The block 'X' represents an obstacle through which the robot cannot travel.

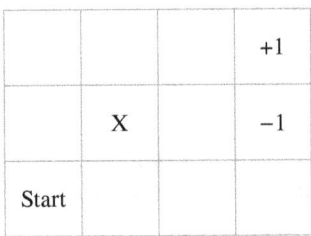

Figure 3.2 A visual representation of Bubbo's world.

In this example, the robot has controls that allow it to move left/right/forward/backward. However, the controls are not reliable, and the robot only moves wherever it intends to with 0.8 probability. With the remaining probability 0.2, the robot moves sideways, that is, with probability 0.1 to each side.

Notice that the Markovian assumption clearly holds for this simple example. What is an optimal solution to this example? A solution concept for a reinforcement learning problem is a mapping from state to action. This mapping is called a "policy" and depends both on the structure of the problem and on the rewards.

For example, if the rewards in all other states are slightly negative, Bubbo wants to reach the terminal state with a persistent +1 reward, as given in Figure 3.3.

→	→	→	+1
↑	X	↑	−1
↑	←	←	←

Figure 3.3 The optimal policy (mapping from state to action) with reward for each step taken being −0.04.

Conversely, if the rewards are very negative, Bubbo just wants to reach any terminal state as soon as possible, as in Figure 3.4.

28   *Markov Decision Processes*

→	→	→	+1
↑	X	→	−1
→	→	→	↑

Figure 3.4 Here, we change the reward of each step to be −2. The optimal policy changes: Now the robot is more concerned with ending the game as quickly as possible rather than trying to get to the +1 reward.

We next describe a formal model to capture this and any other reinforcement learning problem.

3.2 Markov Decision Processes

In this chapter, we describe the general and formal model for reinforcement learning: Markov decision processes.

To ease the exposition, we start with a review of Markov chains, also known as Markov processes, then proceed to more general Markov reward proccesses, and conclude with the full Markov decision process definition.

3.2.1 Markov Chains

A Markov chain (or Markov process) is a model of a memoryless stochastic process as a weighted directed graph. This is the first layer in abstraction before allowing for actions, as it only describes a process without exogenous inputs.

It consists of a set of states and a transition matrix that describes the probability of moving from one state to another. The probabilities of moving between states are the weights on the edges of the graph. The transition matrix is of dimension $S \times S$, where S is the number of states, and in entry i, j we have the probability of moving from state i to state j. Figure 3.5 describes a Markov chain corresponding to a weather process with a corresponding transition matrix P as follows:

$$P = \begin{pmatrix} 0.9 & 0.1 & 0 & 0 \\ 0.5 & 0 & 0.4 & 0.1 \\ 0 & 0.5 & 0 & 0.5 \\ 0 & 0 & 0 & 1 \end{pmatrix}.$$

3.2 Markov Decision Processes

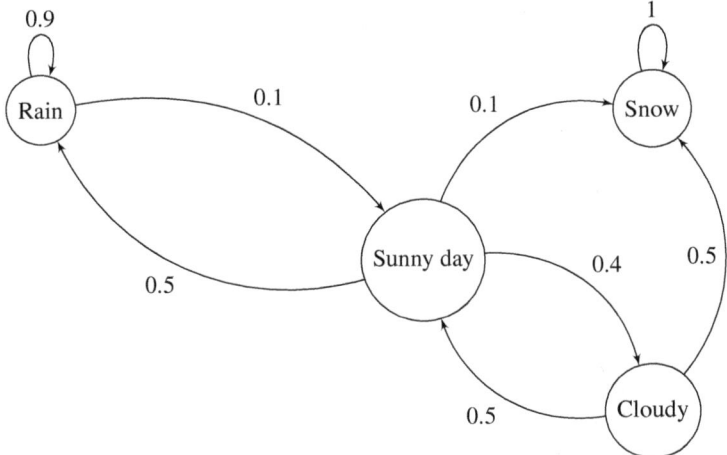

Figure 3.5 A graphical description of a Markov decision process. Each state is the weather for a given day, and the transition probabilities describe the chances of weather types for the next day.

Denote the distribution over states at time t by $\mathbf{p}_t$. Then it can be seen that

$$\mathbf{p}_{t+1} = \mathbf{p}_t P = \mathbf{p}_0 P^{t+1},$$

where $\mathbf{p}_0$ is the initial state distribution represented as a vector.

A central object in the study of Markov chains is stationary distributions.

Definition 3.1 A stationary distribution of a Markov process is a limiting probability distribution

$$\pi = \lim_{t \to \infty} \mathbf{p}_0 P^t.$$

A stationary distribution satisfies

$$\pi = \pi P.$$

A stationary distribution does not always exist nor is it always unique. Existence, uniqueness, and convergence to stationarity are the subject of ergodic theory, a rich and well-studied branch of mathematics. For our discussion, we are interested in sufficient conditions for stationary distributions to exist. They are:

(i) **Irreducibility.** A Markov chain is irreducible if and only if $\forall i, j \; \exists t \; s.t.$ $P_{ij}^t > 0$. In other words, every state is reachable from every other state.

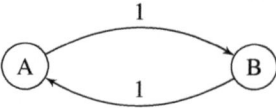

Figure 3.6 In this very simple Markov chain, the state alternates every time step, so there is no unique stationary distribution. This Markov chain has periodicity of two.

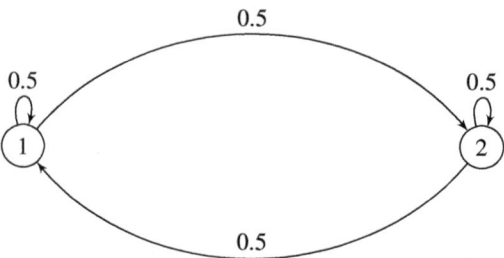

Figure 3.7 Adding a self-loop to every node makes the periodicity one, and this chain is also irreducible. The stationary distribution exists and is unique; it is equal to $\pi = (\frac{1}{2}, \frac{1}{2})$.

(ii) **Aperiodicity.** A Markov chain is aperiodic if and only if $d(s_i) = 1$ for all states $s_i \in S$, where

$$d(s_i) = \gcd\{t \in \mathbb{N}_+ \mid (P^t)_{ii} > 0\},$$

is the periodicity of state s_i, where gcd stands for "greatest common divisor."

Figures 3.6 and 3.7 illustrate the concept of periodicity. The **ergodic theorem** states that every irreducible and aperiodic Markov chain has a unique stationary distribution. Its proof is a beautiful application of spectral graph theory. Exercise 3.1 asks the reader for a guided proof of the ergodic theorem.

Reversible Markov Chains

An interesting special case of Markov chains is the reversible Markov chain.

Definition 3.2 A Markov chain P is said to be reversible with respect to distribution π if and only if the following hold for every two states i, j:

$$\pi_i P_{ij} = \pi_j P_{ji}.$$

A Markov chain is said to be reversible if there exists some distribution with respect to which the Markov chain is reversible.

3.2 Markov Decision Processes

In Exercise 3.4, we prove that a distribution that is reversible for a Markov chain is also stationary for this Markov chain.

3.2.2 Markov Reward Processes

Markov reward processes (MRPs) add the concept of reward to Markov chains. In each state, the agent receives a reward, which, for simplicity, we assume to be in the range $[0, 1]$. In terms of algorithmic consequences, no generality is lost due to this assumption, and proving this is left as an exercise.

The cumulative reward is the sum of discounted rewards over time, so that the immediate rewards are worth more than those received in the future. This is consistent with the economics literature, where a fixed amount of money is worth more today than in the future, because of positive interest rates.

Definition 3.3 A Markov reward process is a tuple (S, P, R, γ), where:

- S is a set of states that can be finite or infinite.
- $P \in \mathbb{R}^{S \times S}$ is the transition matrix.
- $R \in [0, 1]^{S \times S}$ is the reward function, where $R(s, s')$ is the reward for moving from state s to s'.
- $\gamma \in [0, 1]$ is the discount factor.

We henceforth consider instantiations of Markov reward processes, and use the following notation:

(i) We denote by $S_t \in S$ the random state of the process at time t.
(ii) R_t = the random reward received at time t, starting from state S_t and moving to S_{t+1}.
(iii) $\hat{R} \in \mathbb{R}^S$ = the vector that for each state denotes the expected reward in the next iteration as a function of the current state,

$$\hat{R}(s) = \mathbf{E}[R_t | S_t = s].$$

(iv) G_t = the random variable denoting the sum of the discounted rewards starting from time t onwards:

$$G_t = \sum_{i=0}^{\infty} R_{t+i} \gamma^i.$$

When the discount factor is strictly less than one, G_t is well defined. In the case $\gamma = 1$, an alternative definition of G is presented below, assuming that there is a stationary distribution for the underlying Markov chain, denoted by π:

$$G := \lim_{t \to \infty} \frac{1}{t} \sum_{i=1}^{t} R_i = \sum_{s \in S} \pi(s) \hat{R}(s).$$

One of the most important definitions in reinforcement learning is the value function:

Definition 3.4 (Value function) The value function maps states to their expected infinite-horizon reward,

$$\mathbf{v}(s) = \mathbb{E}[G_t | S_t = s].$$

The value function says how valuable a state is as measured by the sum of its future discounted rewards. One way to compute the value function is by using the Bellman equation for the Markov reward process, which is the following recursive equation:

$$\mathbf{v}(s) = \hat{R}(s) + \gamma \sum_{s'} P_{s,s'} \mathbf{v}(s').$$

To see that this equation holds, we can write

$$\mathbf{v}(s) = \mathbb{E}[G_t | S_t = s]$$
$$= \mathbb{E}\left[\sum_{i=0}^{\infty} \gamma^i R_{t+i} | S_t = s\right]$$
$$= \mathbb{E}\left[R_t + \gamma \sum_{i=0}^{\infty} \gamma^i R_{t+1+i} | S_t = s\right]$$
$$= \mathbb{E}[R_t + \gamma G_{t+1} | S_t = s]$$
$$= \mathbb{E}[R_t + \gamma \mathbf{v}(S_{t+1}) | S_t = s]$$
$$= \hat{R}(s) + \gamma \sum_{s'} P_{s,s'} \mathbf{v}(s').$$

In matrix form, we can write the Bellman equation as $\mathbf{v} = \hat{\mathbf{R}} + \gamma P \mathbf{v}$, which simplifies to $\mathbf{v} = (I - \gamma P)^{-1} \hat{\mathbf{R}}$. Given P and $\hat{\mathbf{R}}$, this is a linear system of equations, and thus the value function can be solved using Gaussian elimination in cubic time in the number of states.

3.2.3 Markov Decision Processes

The final layer of abstraction is to allow actions or controls in the language of control theory. This gives the full Markov decision process model as defined below.

3.2 Markov Decision Processes

Definition 3.5 (Markov decision process) A tuple (S, P, R, A, γ) such that:

(i) S is a set of states.
(ii) A is a set of possible actions.
(iii) P is a transition matrix:
$$P^a_{ss'} = \Pr[S_{t+1} = s | S_t = s, A_t = a].$$
(iv) R is a reward function that gives each transition and action a reward $R_{s,s',a}$ which is a reward for moving from s to s' using action a. We henceforth assume without loss of generality that $R \in [0, 1]^{S \times A}$, where we define
$$R_{sa} = \sum_{s' \in S} P_{s,s',a} R_{s,s',a}.$$
(v) $\gamma \in [0, 1)$ is the discount factor.

The central solution concept of study in the theory of reinforcement learning is called a policy. A policy is a mapping from state to distribution over actions, that is,
$$\pi : S \to \Delta(A), \quad \pi(a|s) = \Pr[A_t = a | S_t = s].$$

The goal of reinforcement learning is to determine a sequence of actions to maximize long-term reward. The Markovian structure of the problem suggests that this optimal value is realized by a stationary policy, that is, there exists a fixed map from states to actions that maximizes the expected rewards.

Thus, the goal of reinforcement learning is to find a policy that maximizes the expected reward starting from a certain starting state, that is,
$$\max_{\pi \in S \to \Delta(A)} \mathbf{E} \left\{ \sum_{t=1}^{\infty} \gamma^{t-1} R_t | a_t \sim \pi(s_t), S_1 = s_1 \right\}.$$

where R_t is a random variable that denotes the reward at time t, which depends on the randomness in the policy as well as the transitions.

Similarly to the value function of a Markov reward process, we can define the value function for a Markov decision process as follows.

The value function for a state s given a policy π is defined to be the expected reward from this state onwards where the actions are chosen according to π at each state, that is,
$$\mathbf{v}_\pi(s) = \mathop{\mathbf{E}}_{\pi, P} \left[\sum_{t=1}^{\infty} \gamma^{t-1} R_t | S_1 = s \right].$$

Similarly, the Q function is defined over state–action pairs given a policy π to be the expected cumulative reward gained when starting from some state s

and action a and thereafter choosing actions according to π at each future state, that is,

$$Q_\pi(s,a) = \mathop{\mathbf{E}}_{\pi,P}\left[\sum_{t=1}^\infty \gamma^{t-1} R_t | S_1 = s, A_1 = a\right].$$

A fundamental theorem of reinforcement learning, which is constructively proven in Section 3.4, is as follows.

Theorem 3.6 *There exists a deterministic optimal policy π^*, which maps every state to a single action. All optimal policies achieve the same optimal value $\mathbf{v}^*(s)$ as well as $Q^*(s,a)$, and this holds for all states and actions.*

The fact that there exists an optimal policy that is optimal regardless of the starting state in a MDP is a consequence of the Markovian nature of the transition dynamics.

3.3 The Bellman Equation

A fundamental property of Markov decision processes is that they satisfy the Bellman equation for dynamic programming. The importance of this equation is paramount in algorithm design: We describe two efficient methods for reinforcement learning based on this equation later in this chapter. In later chapters, we also make use of the Bellman equation to derive the stochastic-optimal LQR controller.

The Bellman equation for any policy π relates the value function of that policy to itself through one step of application for that policy. Mathematically, this amounts to the following:

$$\mathbf{v}_\pi(s) = \sum_{a \in A} \pi(a|s)\left[R_{sa} + \gamma \sum_{s' \in S} P^a_{ss'} \mathbf{v}_\pi(s')\right].$$

It is sometimes convenient to use matrix-vector notation, where we can write:

$$\mathbf{v}_\pi = \mathbf{R}^\pi + \gamma P^\pi \mathbf{v}_\pi,$$

where we use the notation

(i) R^π is the expected reward vector induced when playing the policy π, that is, $R^\pi(s) = \mathbf{E}_{a \sim \pi(s)}[R_{sa}]$.
(ii) P^π is the induced transition matrix when playing policy π, that is,

$$P^\pi_{ss'} = \sum_{a \in A} P^a_{ss'} \cdot \pi(a|s).$$

3.3 The Bellman Equation

The Bellman equation gains prominence when applied to the optimal policy of a Markov decision process. Then it is called the **Bellman optimality equation** and states that for every state and action, we have

$$\mathbf{v}^*(s) = \max_a \left\{ R_{sa} + \gamma \sum_{s'} P^a_{ss'} \mathbf{v}^*(s') \right\}.$$

Equivalently, this equation can be written in terms of the Q function as

$$\mathbf{v}^*(s) = \max_a \{Q^*(s, a)\},$$

where Q^* is the Q function of the optimal policy, that is,

$$Q^*(s, a) = R_{sa} + \gamma \sum_{s'} P^a_{ss'} \max_{a'} \{Q^*(s', a')\}.$$

In subsequent chapters, we will also consider cases where the return of interest is either the reward accumulated in a finite number of time steps, that is, $G = \sum_{t=1}^{T} R_t$ in the finite-time setting, or the infinite-time average reward case, that is, $G = \lim_{T \to \infty} \frac{1}{T} \sum_{t=1}^{T} R_t$.

In the finite-time setting, both optimal policies and the optimal value function are nonstationary – they depend on the time step in addition to the current state. In this setting, the corresponding Bellman optimality equation can be written as

$$\mathbf{v}^*_t(s) = \max_a \left\{ R_{sa} + \sum_{s'} P^a_{ss'} \mathbf{v}^*_{t+1}(s') \right\}$$

starting with $\mathbf{v}_T(s) = 0$ for all states.

In the average-reward case, the Bellman optimality equation offers the following characterization: $\mathbf{v}^* : S \to \mathbb{R}$ is the optimal value function if there exists some $\lambda \in \mathbb{R}$ such that

$$\lambda + \mathbf{v}^*(s) = \max_a \left\{ R_{sa} + \sum_{s'} P^a_{ss'} \mathbf{v}^*(s') \right\}.$$

Here the scalar λ corresponds to the optimal average long-term reward, and $\mathbf{v}^*(s)$ reflects the transient advantage of being at state s.

3.3.1 Reinforcement Learning as Linear Programming

One of the immediate conclusions from the Bellman optimality equation is an efficient algorithm for reinforcement learning in Markov decision processes via linear programming.

Consider the problem of computing the value function corresponding to an optimal policy. The optimal value function gives the optimal policy, and deriving this implication explicitly is left as an exercise.

According to the Bellman optimality equation, the optimal value function satisfies

$$\mathbf{v}^*(s) = \max_a \left\{ R_{sa} + \gamma \sum_{s'} P^a_{ss'} \mathbf{v}^*(s') \right\},$$

or in vector form,

$$\mathbf{v}^* = \max_{\mathbf{a} \in A^{|S|}} \{ \mathbf{R}^\mathbf{a} + \gamma P^\mathbf{a} \mathbf{v}^* \},$$

where the maximization over actions takes place independently for each state. Equivalently, we can write the following mathematical program:

$$\min_{\mathbf{v}^*} \mathbf{1}^\top \mathbf{v}^*$$
$$\text{s.t. } \mathbf{v}^*(s) \geq R_{sa} + \gamma \sum_{s'} P^a_{ss'} \mathbf{v}^*(s') \qquad \forall (s,a) \in S \times A.$$

Notice that this is a linear program! The number of variables is $|S|$, and the number of constraints is $|S| \times |A|$, which can be solved in polynomial time in the number of states and actions. The feasibility of this linear program is not immediate but will be established in the next section.

3.4 The Value Iteration Algorithm

In Section 3.3, we established the Bellman equation and its consequence to a polynomial-time solution for reinforcement learning in Markov decision processes. In this section, we study a fundamental algorithm for solving Markov decision processes that is a more direct application of dynamic programming and the Bellman equation.

The value iteration algorithm starts from an arbitrary solution and improves it until a certain measure of optimality is reached. A formal description is given in Algorithm 3.1. The performance guarantee of value iteration, as well as its analysis, is particularly elegant and does not require expertise in mathematical optimization theory.

3.4 The Value Iteration Algorithm

Algorithm 3.1 Value Iteration

Input: MDP
Set $\mathbf{v}_0(s) = 0$ to all states
for $t = 1$ to T **do**
   Update for all states,

$$\mathbf{v}_{t+1}(s) = \max_a \left\{ R_{sa} + \gamma \sum_{s'} P^a_{ss'} \mathbf{v}_t(s') \right\}$$

end for
return π_T which is derived from the last value function as

$$\pi_T(s) = \arg\max_a \left\{ R_{sa} + \gamma \sum_{s'} P^a_{ss'} \mathbf{v}_T(s') \right\}$$

Beyond being simple and providing a finite-time convergence rate, the analysis also establishes the feasibility of the optimal solution, as given in the following theorem.

Theorem 3.7 *Define the residual vector $\mathbf{r}_t(s) = \mathbf{v}^*(s) - \mathbf{v}_t(s)$. Then the optimal value vector $\mathbf{v}^*$ is unique, and the value iteration converges to the optimal value function and policy at a geometric rate of*

$$\|\mathbf{r}_{t+1}\|_\infty \leq \gamma^t \|\mathbf{r}_1\|_\infty.$$

Proof For every state we have by choice of the algorithm that

$$\mathbf{v}_{t+1}(s) = \max_a \left\{ R_{sa} + \gamma \sum_{s'} P^a_{ss'} \mathbf{v}_t(s') \right\} \geq R_{sa^*} + \gamma \sum_{s'} P^{a^*}_{ss'} \mathbf{v}_t(s'),$$

where $a^* = \pi^*(s)$ is the action of an optimal policy at state s. Moreover, from the Bellman optimality equation, we have that

$$\mathbf{v}^*(s) = R_{sa^*} + \gamma \sum_{s'} P^{a^*}_{ss'} \mathbf{v}^*(s').$$

Subtracting the two equations, we get for all states s that

$$\begin{aligned}\mathbf{r}_{t+1}(s) &= \mathbf{v}^*(s) - \mathbf{v}_{t+1}(s) \\ &\leq \gamma \sum_{s'} P^{a^*}_{ss'} (\mathbf{v}^*(s') - \mathbf{v}_t(s')) \\ &= \gamma \sum_{s'} P^{a^*}_{ss'} \mathbf{r}_t(s'),\end{aligned}$$

or in vector form, $\mathbf{r}_{t+1} \leq \gamma P^{\mathbf{a}^*} \mathbf{r}_t$. Furthermore, $\mathbf{r}_t \geq 0$ is entry-wise nonnegative for all t (see Exercise 3.5). Therefore, it holds that

$$\|\mathbf{r}_{t+1}\|_\infty \leq \gamma \|P^{\mathbf{a}^*} \mathbf{r}_t\|_\infty \leq \gamma \|\mathbf{r}_t\|_\infty,$$

where we use the fact that each row of $P^{\mathbf{a}^*}$ sums to one. Applying this repeatedly, we have

$$\|\mathbf{r}_{t+1}\|_\infty \leq \gamma^t \|\mathbf{r}_1\|_\infty.$$

□

3.5 Bibliographic Remarks

The section on reinforcement learning in the text Russel and Norvig (2002) introduces basic concepts, along with commentaries on related topics, such as the origins of the scalar reward model from utility theory. The classical text on reinforcement learning, authored by pioneers of the subject, is Sutton and Barto (2018). In addition, this text extensively discusses early ventures into the field of researchers working in artificial intelligence, psychology, and neuroscience.

Another great reference is the set of lecture notes Silver (2015). See Agarwal et al. (2019a) for an excellent introduction to the theoretical aspects of reinforcement learning, such as sample complexity bounds and reductions to supervised learning. The reader may also refer to Bertsekas (2019) for an extensive survey of computational approaches to learning in unknown Markov decision processes.

The first author had the benefit of a discussion with Arora (2020) on the proof of convergence of value iteration presented in this chapter.

Dynamic programming approaches, in addition to being widely used in algorithm design (Cormen et al., 2022), are at the heart of Markov decision processes. This view is emphasized in Bertsekas (2007), which discusses the sufficiency and necessity of the Bellman optimality conditions at length. It is also a good resource for an involved reader who might be interested in the completeness of the Bellman characterization and the existential concerns for the average-time setting, which we did not expand upon.

Although tabular MDPs do not compactly represent stochastic linear dynamical systems, function-approximation schemes (Du et al., 2021; Jiang et al., 2017; Jin et al., 2020; Kakade et al., 2020) successfully capture the stochastic LQR setting.

3.6 Exercises

3.1. In this exercise, we give a guided proof of the ergodic theorem. Consider an irreducible and aperiodic Markov chain with a transition matrix P.
A. Show that P has a left eigenvalue equal to 1.
B. Show that all eigenvalues of P are at most one in magnitude.
The irreducibility and aperiodicity of the Markov chain imply that the eigenvalue equal to one is unique. We are going to assume this fact and denote by $\mathbf{v}_1$ the corresponding eigenvector.
C. Consider an abirtrary distribution $\mathbf{p}_0$, and the distribution after t iterations of the Markov chain $\mathbf{p}_t = \mathbf{p}_0 P^t$. Prove that $\lim_{t \to \infty} \mathbf{p}_t = \mathbf{v}_1$.

3.2. Prove the geometric convergence of the value iteration algorithm for rewards that can be arbitrary real numbers, as opposed to in the range $[0, 1]$.

3.3. Consider the derivation of an optimal policy for a Markov decision process given its value function as described in Algorithm 3.1. Prove that the resulting policy is indeed optimal.

3.4. Let Markov chain P be reversible with respect to stationary distribution π, that is,
$$\forall i, j, \; \pi_i P_{ij} = \pi_j P_{ji}.$$
Prove that π is stationary for P.

3.5. Prove that the successive iterates $\mathbf{v}_t$ produced by the value iteration algorithm increase monotonically, that is, $\mathbf{v}_{t+1} \geq \mathbf{v}_t$, when initialized with $\mathbf{v}_0 = \mathbf{0}$. Hence, conclude that for any t, $\mathbf{v}^* \geq \mathbf{v}_t$.

3.6. Let P be an irreducible and aperiodic Markov chain. Prove that P is reversible if and only if for any sequence of states $s_1, \ldots s_n$, it holds that
$$P_{s_1 s_2} P_{s_2 s_3} \ldots P_{s_{n-1} s_n} P_{s_n s_1} = P_{s_1 s_n} P_{s_n s_{n-1}} \ldots P_{s_2 s_1}.$$

4
Linear Dynamical Systems

In Chapter 2, we observed that even simple questions about general dynamical systems are intractable. This motivates the restriction to a special class of dynamical systems for which we can design efficient algorithms with provable guarantees.

Our treatment in this chapter is restricted to linear dynamical systems (LDS): dynamical systems that evolve linearly according to the equation below (see schematic depiction in Figure 4.1):

$$\mathbf{x}_{t+1} = A_t \mathbf{x}_t + B_t \mathbf{u}_t + \mathbf{w}_t.$$

If the matrices $A_t = A$ and $B_t = B$ are fixed, we say that the dynamical system is linear time-invariant (LTI). We have seen examples of LTI systems in Chapter 2, namely the double integrator and the aircraft dynamics.

A more general class of linear dynamical systems is when the matrices A_t and B_t change over time. These are called linear time-varying (LTV) systems. The class of LTV systems can express more general dynamics via linearization, as we show next.

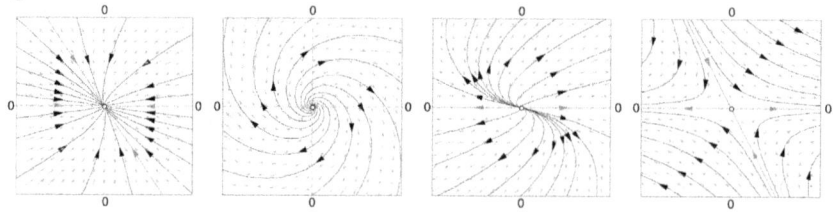

Figure 4.1 Four different deterministic linear time-invariant dynamical systems represented via their induced vector fields and a few chosen trajectories. Created by XaosBits using the Mathematica packages CurvesGraphics.nb by Gianluca Gorni and DrawGraphics.nb by David and Alice Park, licensed under CC-BY-2.5.

4.1 Approximating General Dynamics via LTV Systems

Consider any dynamics that can be expressed by a smooth function f. Such a system can be approximated by a time-varying linear system to an arbitrary accuracy determined by the time-step interval between samples.

To see this, first consider the one-dimensional first-order Taylor expansion of a given dynamics function f, which is twice differentiable, around an anchor point $(\bar{x}_t, \bar{u}_t)$:

$$x_{t+1} = f(x_t, u_t) + w_t$$
$$= f(\bar{x}_t, \bar{u}_t) + \underbrace{\frac{\partial}{\partial x} f(x, u)\bigg|_{\bar{x}_t, \bar{u}_t}}_{\stackrel{\text{def}}{=} A_t} (x_t - \bar{x}_t) + \underbrace{\frac{\partial}{\partial u} f(x, u)\bigg|_{\bar{x}_t, \bar{u}_t}}_{\stackrel{\text{def}}{=} B_t} (u_t - \bar{u}_t) + w_t,$$

where w_t is the approximation error, or remainder, which by the mean value theorem behaves as $w_t \sim (x_t - \bar{x}_t)^2 + (u - \bar{u}_t)^2$. Thus, this simplifies to

$$x_{t+1} = A_t x_t + B_t u_t + w_t.$$

An analogous approximation holds for the higher-dimensional case. The Taylor expansion is

$$f(\mathbf{x}, \mathbf{u}) = f(\bar{\mathbf{x}}, \bar{\mathbf{u}}) + J_\mathbf{x} f(\bar{\mathbf{x}}, \bar{\mathbf{u}})(\mathbf{x} - \bar{\mathbf{x}}) + J_\mathbf{u} f(\bar{\mathbf{x}}, \bar{\mathbf{u}})(\mathbf{u} - \bar{\mathbf{u}}) + w_t,$$

where $J_\mathbf{x} f(\bar{\mathbf{x}}, \bar{\mathbf{u}})$ is the Jacobian of f with respect to variables $\bar{\mathbf{x}}$. Writing the matrices A_t, B_t for the Jacobians, we have

$$\mathbf{x}_{t+1} = A_t \mathbf{x}_t + B_t \mathbf{u}_t + \mathbf{w}_t.$$

This derivation explains why in practice the class of time-varying linear dynamical systems is very expressive, at least for smooth dynamical systems. This is particularly true with hidden states that can capture complicated dynamics while imposing enough structure for efficient algorithms.

As we shall see next, stabilizability and controllability can be checked efficiently for linear systems, in contrast to general dynamics.

4.2 Stabilizability of Linear Systems

Verifying stability of a system, and more so stabilizability, was shown to be intractable in our discussion of Chapter 2. For linear dynamical systems, however, the picture is quite different: Stability and stabilizability can be efficiently verified! Furthermore, the theoretical analysis of this fact is interesting and elegant.

For simplicity, in the rest of this section we consider time-invariant linear dynamical systems of the form

$$\mathbf{x}_{t+1} = A\mathbf{x}_t + B\mathbf{u}_t + \mathbf{w}_t,$$

although much of our discussion can be generalized to LTV systems.

Recall that a system is stabilizable if, from every starting state, there is a sequence of controls that drives the system to zero. Determining whether a linear time-invariant system is stabilizable is determined by a particularly simple condition as follows. Henceforth, we denote by $\rho(A)$ the spectral radius of a matrix A, that is,

$$\rho(A) = \max\{|\lambda_1|, \ldots, |\lambda_d|\},$$

the maximum modulus of all (possibly complex) eigenvalues. The stabilizability of a linear dynamical system can be characterized as follows.

Theorem 4.1 *A time-invariant LDS is stabilizable if and only if there exists a matrix K such that*

$$\rho(A + BK) < 1.$$

We therefore prove the sufficiency of this condition and leave the necessity as Exercise 4.1. Both directions rely on the characterization of the spectral norm as follows:

Fact 4.2 $\rho(A) < 1$ *if and only if* $\lim_{t \mapsto \infty} A^t = 0$. *Furthermore, if* $\rho(A) > 1$, *then for any norm,* $\lim_{t \mapsto \infty} \|A^t\| = \infty$.

The proof of this fact is left as Exercise 4.2. Based on this fact, we can now prove the sufficiency part of Theorem 4.1.

Theorem 4.1, sufficiency Suppose that there exists a K such that $\rho(A+BK) < 1$. Then choose all controls to be $\mathbf{u}_t = K\mathbf{x}_t$. The evolution of the system without perturbation is given by

$$\mathbf{x}_t = (A + BK)\mathbf{x}_{t-1} = \cdots = (A + BK)^t \mathbf{x}_0.$$

Thus, using the previous fact concerning the spectral radius, we have

$$\lim_{t \to \infty} \|\mathbf{x}_t\| = \lim_{t \to \infty} \|(A + BK)^t \mathbf{x}_0\| = \|0\mathbf{x}_0\| = 0.$$

□

4.2.1 Efficient Verification of Stabilizability

The spectral radius of a matrix can be computed to an arbitrary accuracy in polynomial time and can be approximated by the power method. This implies that the stability of a given system can be efficiently verified.

To efficiently check if a given system is stabilizable, we make note of the following result due to Lyapunov. We use the notation $A \succeq 0$ to denote that the matrix A is positive semidefinite, meaning that it is symmetric and all its eigenvalues are nonnegative.

Theorem 4.3 $\rho(A) < 1$ *holds if and only if there exists a matrix* $P \in \mathbb{R}^{d_x \times d_x}$ *such that* $P \succeq I$ *and* $P \succ A^\top P A$.

Proof Say there exists a matrix P such that $P \succeq I$ and $P \succ A^\top P A$; then to show that $\rho(A) < 1$, it suffices to prove that $\lim_{t \to \infty} A^t = 0$. To observe the latter, first note that there exists a positive constant $c < 1$ such that $cP \succeq A^\top P A$, implying that $c^t P \succeq (A^t)^\top P A^t \succeq (A^t)^\top A^t$ via repeated applications of the previous inequality. Therefore, we have $\lim_{t \to \infty} (A^t)^\top A^t = 0$, and consequently $\lim_{t \to \infty} A^t = 0$ since $\|A^\top\| \|A\| = \|A^\top A\|$.

For the other direction, given some matrix A with $\rho(A) < 1$, consider $P = \sum_{t=0}^{\infty} (A^t)^\top A^t$. Since $P = I + A^\top P A$, P satisfies the required conclusions. The fact that P as defined exists, that is, that it is finite, as long as $\rho(A) < 1$ can be verified via the Jordan normal form of A, which in particular asserts that $\|A^t\| \leq O(t^{d_x} \rho(A)^t)$. □

Using this observation, one can write a semidefinite mathematical program that checks if a given system is stabilizable and if so produces a matrix K that guarantees a spectral radius smaller than one.

Theorem 4.4 *There exists a matrix K such that $\rho(A + BK) < 1$ if and only if there exist matrices $P \in \mathbb{S}^{d_x \times d_x}$ and $L \in \mathbb{R}^{d_u \times d_x}$ such that*

$$\begin{bmatrix} P & (AP + BL)^\top \\ AP + BL & P \end{bmatrix} \succ 0, \qquad P \succeq I. \tag{4.1}$$

Furthermore, given such matrices P and L satisfying Equation 4.1, the matrix $K' = LP^{-1}$ satisfies $\rho(A + BK') < 1$.

Proof We begin by observing the following result, using the Schur complement, regarding the semi-definiteness of block matrices.

Lemma 4.5 *Define a block matrix M as*

$$M = \begin{bmatrix} A & B \\ B^\top & C \end{bmatrix}. \tag{4.2}$$

Then, $M \succ 0$ if and only if $A \succ 0$ and $C - B^\top A^{-1} B \succ 0$.

Note that since the spectral radius purely involves the eigenvalues of a matrix, it is invariant under transposition, and we have $\rho((A + BK)^\top) = \rho(A + BK)$. Now using this fact and Theorem 4.3, we know that the stabilizability of a linear system is equivalent to the existence of matrices K and P such that $P \succeq I$ and $P \succ (A + BK)P(A + BK)^\top$. The above lemma allows us to reformulate this cubic constraint to a quadratic constraint. The stabilizability of a system is equivalent to existence of matrices K and P such that

$$\begin{bmatrix} P & ((A+BK)P)^\top \\ (A+BK)P & P \end{bmatrix} \succ 0, \qquad P \succeq I.$$

Lastly, we reparameterize this constraint as a linear semidefinite constraint as demonstrated below, by choosing $L = KP$.

$$\begin{bmatrix} P & (AP+BL)^\top \\ AP+BL & P \end{bmatrix} \succ 0, \qquad P \succeq I.$$

Notice that this is indeed a reformulation of the mathematical feasibility program, and not a relaxation, since the reparamertization $K = LP^{-1}$ is invertible. □

4.3 Controllability of Linear Dynamical Systems

Recall that a system is controllable if and only if in the absence of noise and for every target state there is a sequence of controls that drives the system state to that target. Linear dynamical systems exhibit a particularly elegant characterization of controllability. First, we need to define an object of central importance for linear dynamical systems, the Kalman matrix.

Definition 4.6 The *Kalman controllability matrix* of a linear dynamical system is given by

$$K_r(A, B) = [B, AB, A^2B, \ldots, A^{r-1}B] \in \mathbb{R}^{d_x \times r d_u}.$$

The controllability of a linear dynamical system can now be characterized using this matrix. Recall that the rank of a matrix is the maximal number of linearly independent columns (or rows).

Theorem 4.7 *A time-invariant linear dynamical system is controllable if and only if*

$$\operatorname{rank} K_{d_x}(A, B) = d_x.$$

4.3 Controllability of Linear Dynamical Systems

Proof Consider the evolution of the noiseless linear dynamical system starting from the initial state $\mathbf{x}_0$. The state at time T is given by the expression

$$\mathbf{x}_T = A\mathbf{x}_{T-1} + B\mathbf{u}_{T-1}$$
$$= A^2\mathbf{x}_{T-2} + AB\mathbf{u}_{T-2} + B\mathbf{u}_{T-1}$$
$$= \cdots = \sum_{t=1}^{T} A^{t-1} B\mathbf{u}_{T-t} + A^T \mathbf{x}_0.$$

Thus,

$$\mathbf{x}_T - A^T \mathbf{x}_0 = [B, AB, A^2 B, \ldots, A^{T-1} B] \begin{bmatrix} \mathbf{u}_{T-1} \\ \vdots \\ \mathbf{u}_0 \end{bmatrix}.$$

By the fundamental theory of linear algebra, this equation has a solution if and only if $K_T(A, B)$ has full rank. That is, for every initial state $\mathbf{x}_0$ and final state $\mathbf{x}_T$, there exists a set of control vectors $\{\mathbf{u}_t\}$ that satisfy the equation if and only if $K_T(A, B)$ has full rank. The rank of the Kalman matrix is upper bounded by the number of rows of K_T, which is d_x. Thus, the system is controllable if and only if the rank of K_T is d_x for all $T \geq d_x$.

Denote by $r_t = \text{rank}(K_t)$ the rank of the Kalman matrix after t iterations. Since the rank can never decrease with t, it remains to show that the system is controllable if and only if $r_t = r_{d_x}$ for all $t \geq d_x$.

We claim that once $r_t = r_{t+1}$, the rank never increases again by increasing t. Assuming $r_t = r_{t+1}$, we have

$$\text{span}[B, AB, \ldots, A^t B] = \text{span}[B, AB, \ldots, A^{t-1} B].$$

Since linear transformations retain containment, we can multiply by A and get

$$\text{span}[AB, \ldots, A^{t+1} B] \subseteq \text{span}[AB, \ldots, A^t B].$$

Adding B to both matrices does not affect the subset relationship:

$$\text{span}[B, AB, \ldots, A^{t+1} B] \subseteq \text{span}[B, AB, \ldots, A^t B].$$

Therefore, $r_{t+2} \leq r_{t+1}$. Since the rank cannot decrease through the addition of columns, we have $r_{t+2} = r_{t+1}$. Since the rank can grow at most d_x times, equation $r_{d_x} = r_T$ holds for all $T > d_x$ if and only if the system is controllable, and this concludes the proof. □

4.4 Quantitative Definitions

In the rest of this text our goal is to give efficient algorithms with finite time performance bounds for online control. These finite-time bounds naturally depend on how difficult it is for a given system to stabilize. Therefore, we require a quantitative definition of stabilizability, one that holds for both time-invariant and time-varying linear dynamical systems.

4.4.1 Stabilizability of Time-Invariant Systems

Consider a linear time-invariant system given by the transformations (A, B). We have already remarked that this system is called stabilizable, according to the classical definition, if and only if there exists K such that $\rho(A + BK) < 1$ holds.

A stronger quantitative property can be obtained as follows, and in fact it is implied by BIBO stabilizability as defined in Definition 2.4. The proof of this fact is left as Exercise 4.5.

Lemma 4.8 *Consider a linear controller K for the LTI system (A, B). If there exist $\kappa > 0$ and $\delta \in (0, 1]$ such that the matrix $A + BK$ admits the following inequality for all t, then K is a γ-stabilizing controller for (A, B).*

$$\|(A + BK)^t\| \leq \kappa(1 - \delta)^t, \quad \frac{\kappa}{\delta} \leq \gamma.$$

Proof Let $\tilde{K} = A + BK$. By unrolling the dynamics equation, we have that

$$\mathbf{x}_{t+1} = A\mathbf{x}_t + B\mathbf{u}_t + \mathbf{w}_t = \tilde{K}\mathbf{x}_t + \mathbf{w}_t$$

$$= \sum_{i=0}^{t-1} \tilde{K}^i \mathbf{w}_{t-i} + \tilde{K}^{t+1}\mathbf{x}_0.$$

Thus, as long as $\|\mathbf{w}_t\|, \|\mathbf{x}_0\| \leq 1$, we have

$$\|\mathbf{x}_{t+1}\| \leq \sum_{i=0}^{t} \|\tilde{K}^i\| \|\mathbf{w}_{t-i}\| + \|\tilde{K}^{t+1}\| \|\mathbf{x}_0\|$$

$$\leq \kappa \sum_{i=0}^{\infty} (1 - \delta)^i \leq \frac{\kappa}{\delta} \leq \gamma.$$

□

A linear time-invariant system is thus γ-stabilizable if we can find a matrix K with the above properties.

4.4.2 Stabilizability of Time-Varying Systems

The γ-stabilizability condition, as opposed to classical stabilizability, is particularly important to generalize to time-varying systems, since the spectral radius of the product is **not** bounded by the product of spectral radii; see Exercise 4.3.

We now define a sufficient condition for γ-stabilizability. Consider a time-varying linear dynamical system given by the transformations $\{A_t, B_t\}$.

Lemma 4.9 *Consider a sequence of linear controllers $K_{1:T}$ for the time-varying linear dynamical system $\{A_{1:T}, B_{1:T}\}$. If the matrices $\tilde{K}_t = A_t + B_t K_t$ satisfy the following inequality, then this linear controller sequence is γ-stabilizing.*

$$\forall t_1, t_2 \ . \ \left\| \prod_{t=t_1}^{t_2} \tilde{K}_t \right\| \leq \kappa(1-\delta)^{t_1-t_2}, \ \frac{\kappa}{\delta} \leq \gamma.$$

Proof Notice that under the condition of the lemma, by unrolling the dynamics equation we have that

$$\|\mathbf{x}_{t+1}\| = \|A_t \mathbf{x}_t + B_t \mathbf{u}_t + \mathbf{w}_t\| = \|\tilde{K}_t \mathbf{x}_t + \mathbf{w}_t\|$$

$$= \left\| \sum_{i=0}^{t-1} (\prod_{j=1}^{i} \tilde{K}_{t-j}) \mathbf{w}_{t-i} + (\prod_{j=1}^{t} \tilde{K}_{t-j}) \mathbf{x}_0 \right\|$$

$$\leq \kappa \sum_{i=0}^{\infty} e^{-\delta i} \leq \frac{\kappa}{\delta} \leq \gamma.$$

□

4.4.3 Controllability of Time-Invariant Systems

Consider the following one-dimensional dynamical system for some small $\varepsilon > 0$:

$$x_{t+1} = x_t + \varepsilon u_t.$$

Clearly, such a system is nominally controllable, as indeed is any one-dimensional system with nonzero B. However, reaching a state $x = 1$ from an initial $x_0 = 0$ in a constant number of time steps requires a large control input u_t of magnitude at least $1/\varepsilon$. Thus, such a system may be said to be barely controllable. This difficulty of controllability is captured in the definition below, which requires not only that $K(A, B)$ is a full row-rank but also that such a statement is robust to small changes in A, B.

Definition 4.10 A time-invariant linear dynamical system (A, B) is (k, κ)-strongly controllable if and only if

$$\text{rank } K_k(A, B) = d_x, \text{ and } \|(K_k(A, B)K_k(A, B)^\top)^{-1}\| \leq \kappa.$$

In the above definition, in addition to the full row-rank constraint, the spectral norm of the inverse of the Gram matrix of $K(A, B)$ is bounded by κ. For example, in the simple one-dimensional system we presented above, with $k = 1$, the matrix $K(A, B)$ is ε, and thus $\kappa = \frac{1}{\varepsilon^2}$, which indeed captures the complexity of controlling the system.

4.5 Bibliographic Remarks

The idea of reducing a nonlinear dynamical system to a time-varying linear dynamical system had been independently discovered in disparate fields. In the context of control, using the linearization to characterize local stability properties of nonlinear systems was pioneered by Lyapunov (1992) and is called the Lyapunov direct method; see, for example, Slotine and Li (1991).

The asymptotic notions of stability, stabilizability, and controllability discussed in this chapter constitute a core part of the theory of linear control and are found in most texts on the subject (e.g. Hespanha (2018); Stengel (1994)).

The efficient verification of stabilizability via a semidefinite program and the recovery of a stable controller form the solution of the latter appears in Ahmadi (2016). However, it is known that even imposition of norm constraints on a plausibly stable controller makes this problem computationally hard (Blondel and Tsitsiklis, 1997, 2000, 1999). The somewhat distinct but related problems of stabilizability with even a one-dimensional linear constraint remain open to date.

The notion of γ-stabilizability has a precursor in the notion of strong stability introduced in Cohen et al. (2018). An extension of this strong stability concept to time-varying linear dynamics was established in Cohen et al. (2019). The definition presented in this chapter has weaker requirements than those in the listed articles, allowing us to work with and guarantee regret later with respect to a larger class of comparator policies.

4.6 Exercises

4.1. * Prove that the existence of a matrix K such that $\rho(A + BK) < 1$ is a necessary condition for a linear dynamical system to be stabilizable. Hint: this exercise requires the optimal solution to LQR in the next chapter.

4.6 Exercises

4.2. Prove that $\rho(A) < 1$ if and only if $\lim_{t \to \infty} A^t = 0$. Furthermore, if $\rho(A) > 1$, then for any norm, $\lim_{t \to \infty} \|A^t\| = \infty$. Finally, prove that $\rho(A) = \lim_{t \to \infty} \|A^t\|^{1/t}$.

4.3. Prove that the spectral radius of a product of matrices is not necessarily bounded by the product of the spectral radii of the individual matrices. That is, show that in general,

$$\rho\left(\prod_{i=1}^{n} A_i\right) \nleq \prod_{i=1}^{n} \rho(A_i).$$

4.4. Prove that a stochastic linear dynamical system with Gaussian perturbations

$$\mathbf{x}_{t+1} = A\mathbf{x}_t + \mathbf{w}_t, \qquad \mathbf{w}_t \sim \mathcal{N}(0, I),$$

constitutes a reversible Markov chain if and only if A is symmetric, that is, $A = A^\top$.
Hint: One way to do this exercise is to explicitly calculate the stationary distribution of an LDS and then to verify reversibility as stated in its definition. Exercise 3.6 provides an alternative without the need to derive the stationary distribution.

4.5. This exercise explores the consequences of γ-BIBO stability (Definition 2.4) for linear systems. Although we have already seen sufficient conditions to ensure BIBO stability, we ask for the necessity of such conditions.
A. For a time-invariant linear system (A, B) equipped with a linear policy K, prove that BIBO stability indeed implies $\rho(A + BK) < 1$. In fact, comment on how large $\rho(A + BK)$ can be.
B. For a time-invariant BIBO-stable linear system, prove that there exist $\kappa > 0$ and $\delta \in (0, 1]$ such that, for all t, $\|(A + BK)^t\| \leq \kappa(1 - \delta)^t$. In particular, this proves that the conditions listed in Lemma 4.8 are necessary, in addition to being sufficient.
C. Construct a time-varying linear system that is BIBO stable but does not have the exponential decay properties required in Lemma 4.9.

4.6. Prove that the bound of one on the operator norm of the matrix A is a sufficient but not necessary condition for stabilizability.

5
Optimal Control of Linear Dynamical Systems

In the chapters thus far, we have studied more basic notions than control, namely solution concepts of dynamical systems such as stabilizability and controllability. Even for these more basic problems, we have seen that they are computationally intractable in general.

This motivated us to look into simpler dynamics, namely linear dynamical systems, for which we can at least answer basic questions about stabilizability and controllability. In this chapter, we consider how we can control linear dynamical systems optimally.

The answer is positive in a strong sense. An archetypal problem in optimal control theory is the control of linear dynamical systems with quadratic cost functions. This setting is called the *linear quadratic regulator*, which we define and study in this chapter. The resulting efficient and practical algorithm is one of the most widely used subroutines in control.

5.1 The Linear Quadratic Regulator

The linear quadratic regulator (LQR) problem is that of optimal control of a linear dynamical system with known and fixed quadratic costs. For simplicity, we mostly consider the time-invariant version, where the linear dynamical system is fixed, and the costs are also fixed quadratic, as given formally below:

Definition 5.1 The (time-invariant) linear quadratic regulator (LQR) input is a linear dynamical system that evolves according to

$$\mathbf{x}_{t+1} = A\mathbf{x}_t + B\mathbf{u}_t + \mathbf{w}_t,$$

where $\mathbf{w}_t \sim \mathcal{N}(0, \sigma^2)$ is i.i.d. Gaussian, and convex quadratic cost functions are specified by matrices Q and R.

5.2 Optimal Solution of the LQR

The goal is to generate a sequence of controls $\mathbf{u}_t$ that minimizes the expected cost, taken over the randomness of the perturbations, given by quadratic loss functions,

$$\min_{\mathbf{u}(\mathbf{x})} \mathbb{E}\left[\sum_{t=1}^{T} \mathbf{x}_t^\top Q \mathbf{x}_t + \mathbf{u}_t^\top R \mathbf{u}_t\right].$$

To remind the reader, the notation we use is as follows.

(i) $\mathbf{x}_t \in \mathbb{R}^{d_x}$ is the state.
(ii) $\mathbf{u}_t \in \mathbb{R}^{d_u}$ is the control input.
(iii) $\mathbf{w}_t \in \mathbb{R}^{d_w}$ is the perturbation sequence drawn i.i.d. from normal distributions, or any other zero-mean distribution as long as the draws are independent of states or controls.
(iv) $A \in \mathbb{R}^{d_x \times d_x}$, $B \in \mathbb{R}^{d_x \times d_u}$ are the system matrices.
(v) $Q \in \mathbb{R}^{d_x \times d_x}$, $R \in \mathbb{R}^{d_u \times d_u}$ are positive definite matrices. Thus, the cost functions are convex.

This framework is very useful, since many physical problems, such as the centrifugal governor and the ventilator control problem, can be approximated by a linear dynamical system. In addition, quadratic convex cost functions are natural in many physical applications.

The LQR is a special case of a Markov decision process, which we described in Chapter 3, since the transitions are stochastic. However, we cannot naively apply value iteration, policy iteration, or linear programming since there are infinitely many states and actions.

We proceed to describe a particularly elegant solution to the LQR problem that makes use of its special structure.

5.2 Optimal Solution of the LQR

We know from Chapter 3 that the LQR is a special case of a finite-time Markov decision process and hence admits an optimal policy. However, the optimal policy is in general a mapping from states to actions, both of which are infinite Euclidean spaces in our case.

However, it can be shown that the optimal policy for an LQR is a **linear policy**, that is, a linear mapping from $\mathbb{R}^{d_x}$ to $\mathbb{R}^{d_u}$! This astonishing fact, together with a remarkable property of the value function, is formally given in the following theorem. We state the theorem for a finite number of iterations and for a linear time-invariant system. Generalizing this to a time-varying linear system is left as Exercise 5.1.

Theorem 5.2 *For the LQR problem with finite $T > 0$ iterations, there exist positive semi-definite matrices S_t, K_t, and scalars c_t, such that the optimal value function and optimal policy satisfy:*

$$\mathbf{v}_t^*(\mathbf{x}) = \mathbf{x}^\top S_t \mathbf{x} + c_t, \quad \mathbf{u}_t^*(\mathbf{x}) = K_t \mathbf{x}.$$

Proof Recall that the Bellman optimality equation for a general MDP is written as

$$\mathbf{v}_t^*(s) = \max_a \left\{ R_{sa} + \sum_{s'} P_{ss'}^a \mathbf{v}_{t+1}^*(s') \right\}.$$

We henceforth write the special case applicable to LQR and prove the theorem by backward induction on t.

Base Case: For $t = T$, we have

$$\mathbf{v}_T^*(\mathbf{x}) = \min_{\mathbf{u}} \left\{ \mathbf{x}^\top Q \mathbf{x} + \mathbf{u}^\top R \mathbf{u} \right\}.$$

Since $R \succ 0$, a minimum occurs at $\mathbf{u} = 0$. Hence, $S_T = Q \succ 0$, $K_T = 0$, and $c_T = 0$.

Inductive Step: Assume for some $t > 1$ that $\mathbf{u}_t^*(\mathbf{x}) = K_t \mathbf{x}$ and $\mathbf{v}_t^*(\mathbf{x}) = \mathbf{x}^\top S_t \mathbf{x} + c_t$. Then we have

$$\mathbf{v}_{t-1}^*(\mathbf{x}) = \min_{\mathbf{u}} \left\{ \mathbf{x}^\top Q \mathbf{x} + \mathbf{u}^\top R \mathbf{u} + \mathop{\mathbf{E}}_{\mathbf{w}} \mathbf{v}_t^*(A\mathbf{x} + B\mathbf{u} + \mathbf{w}) \right\}$$

$$= \min_{\mathbf{u}} \left\{ \mathbf{x}^\top Q \mathbf{x} + \mathbf{u}^\top R \mathbf{u} + \mathop{\mathbf{E}}_{\mathbf{w}} \left[(A\mathbf{x} + B\mathbf{u} + \mathbf{w})^\top S_t (A\mathbf{x} + B\mathbf{u} + \mathbf{w}) \right] \right\} + c_t$$

$$= \min_{\mathbf{u}} \left\{ \mathbf{x}^\top Q \mathbf{x} + \mathbf{u}^\top R \mathbf{u} + (A\mathbf{x} + B\mathbf{u})^\top S_t (A\mathbf{x} + B\mathbf{u}) + \mathop{\mathbf{E}}_{\mathbf{w}} [\mathbf{w}^\top S_t \mathbf{w}] \right\} + c_t,$$

where in the last equality we used the fact that $\mathbf{E}[\mathbf{w}] = 0$, and that the perturbation is independent of the state and control. We henceforth denote the second moment of the noise by $\mathbf{E}[\mathbf{w}\mathbf{w}^\top] = \sigma^2 I$. This implies that for any matrix M that is independent of $\mathbf{w}$, we have $\mathbf{E}[\mathbf{w}^\top M \mathbf{w}] = \mathbf{E}[M \mathbf{w}\mathbf{w}^\top] = \sigma^2 \cdot M \bullet I = \sigma^2 \mathrm{Tr}(M)$.

Since R and S_t are positive semidefinite, the above expression is convex as a function of $\mathbf{u}$. At the global minimum the gradient vanishes, and hence

$$\nabla_\mathbf{u} \left[\mathbf{x}^\top Q \mathbf{x} + \mathbf{u}^\top R \mathbf{u} + (A\mathbf{x} + B\mathbf{u})^\top S_t (A\mathbf{x} + B\mathbf{u}) + \sigma^2 \mathrm{Tr}(S_t) \right] = 0.$$

Since Q, R, S_t are all positive semi-definite and symmetric, this gives us

$$2R\mathbf{u} + 2B^\top S_t A\mathbf{x} + 2B^\top S_t B\mathbf{u} = 0 \implies (R + B^\top S_t B)\mathbf{u} = -B^\top S_t A\mathbf{x}.$$

Hence, we have that

$$\mathbf{u}_{t-1}(\mathbf{x}) = K_{t-1}\mathbf{x}, \quad K_{t-1} = -(R + B^\top S_t B)^{-1}(B^\top S_t A).$$

Notice that the matrix $R + B^\top S_t B$ is full rank since $R \succ 0$. Plugging this into the expression for the value function, we get

$$\begin{aligned}\mathbf{v}^*_{t-1}(\mathbf{x}) &= \mathbf{x}^\top Q\mathbf{x} + \mathbf{x}^\top K_{t-1}^\top R K_{t-1}\mathbf{x} + \mathbf{x}^\top (A - BK_{t-1})^\top S_t(A - BK_{t-1})\mathbf{x} + \sigma^2 \mathbf{Tr}(S_t) + c_t \\ &= \mathbf{x}^\top S_{t-1}\mathbf{x} + c_{t-1},\end{aligned}$$

where $S_{t-1} = Q + K_{t-1}^\top R K_{t-1} + (A - BK_{t-1})^\top S_t (A - BK_{t-1})$ is positive definite, and $c_{t-1} = c_t + \sigma^2 \mathbf{Tr}(S_t)$.

$\square$

5.3 Infinite-Horizon LQR

Although infinite-horizon problems can never appear in the real world, in practice the horizon length may be unknown. Although there are techniques to iteratively update the horizon length with geometrically increasing guesstimates, these are slow and impractical.

An alternative is to solve the LQR for the infinite-horizon case, which results in a very elegant solution. Not only that; this turns out to be a practical solution to which the finite-horizon solution converges quickly.

The infinite-horizon LQR problem can be written as follows:

$$\min_{\mathbf{u}(\mathbf{x})} \lim_{T \to \infty} \frac{1}{T} \mathbb{E}\left[\sum_{t=1}^T \mathbf{x}_t^\top Q \mathbf{x}_t + \mathbf{u}_t^\top R \mathbf{u}_t\right]$$
$$\text{s.t. } \mathbf{x}_{t+1} = A\mathbf{x}_t + B\mathbf{u}_t + \mathbf{w}_t,$$

where, similar to the finite-horizon case, we assume that the perturbations $\mathbf{w}_t$ are i.i.d. zero mean and have bounded variance. We henceforth assume that the perturbations are all zero for simplicity, and adding zero-mean bounded-variance perturbations is left as Exercise 5.2.

Without perturbations, the Bellman optimality equation becomes

$$\mathbf{v}^*(\mathbf{x}) = \min_{\mathbf{u}} \left\{ \mathbf{x}^\top Q \mathbf{x} + \mathbf{u}^\top R \mathbf{u} + \mathbf{v}^*(A\mathbf{x} + B\mathbf{u}) \right\}.$$

Assuming that the value function is quadratic, we have:

$$\mathbf{x}^\top S \mathbf{x} = \min_{\mathbf{u}} \left\{ \mathbf{x}^\top Q \mathbf{x} + \mathbf{u}^\top R \mathbf{u} + (A\mathbf{x} + B\mathbf{u})^\top S (A\mathbf{x} + B\mathbf{u}) \right\}.$$

Similarly to the finite-time case, we have the following result.

Lemma 5.3 *The optimal policy is given by a transformation* $\mathbf{u}^*(\mathbf{x}) = K\mathbf{x}$.

Proof At optimality, the gradient with respect to **u** of the above equation is zero. This implies that

$$2R\mathbf{u} + 2B^{\mathsf{T}}SB\mathbf{u} + 2B^{\mathsf{T}}SA\mathbf{x} = 0,$$

which implies

$$K = -(R + B^{\mathsf{T}}SB)^{-1}(B^{\mathsf{T}}SA).$$

$\square$

Plugging this back into the Bellman equation, we obtain

$$\mathbf{x}^{\mathsf{T}}S\mathbf{x} = \mathbf{x}^{\mathsf{T}}Q\mathbf{x} + \mathbf{x}^{\mathsf{T}}K^{\mathsf{T}}RK\mathbf{x} + \mathbf{x}^{\mathsf{T}}(A + BK)^{\mathsf{T}}S(A + BK)\mathbf{x}.$$

Since this equality holds for all **x**, we get

$$S = Q + K^{\mathsf{T}}RK + (A + BK)^{\mathsf{T}}S(A + BK).$$

Simplifying this expression, we obtain

$$S = Q + A^T SA - A^T SB\left(R + B^T SB\right)^{-1} B^T SA.$$

This equation is called the **discrete time algebraic Riccati equation** (DARE) and is one of the most important equations in control theory.

5.3.1 Solving the DARE

In the one-dimensional case, although the right side of the equation is a rational function, we can multiply both sides by $R + B^{\mathsf{T}}SB$ to form a quadratic equation in S. Solving the DARE therefore corresponds to finding a nonnegative root of a quadratic polynomial in the one-dimensional case. The reason we only consider the nonnegative root is that the value function is nonnegative by our choice of quadratic convex costs.

In higher dimensions, the correct generalization is somewhat more difficult to see, but using Lemma 4.5, the equation can be reformulated as the solution to the following semidefinite program:

$$\min_{X \in S^{d_x \times d_x}} \langle X, I \rangle$$

$$\text{s.t.} \begin{bmatrix} X & B^T SA \\ A^T SB & R + B^T SB \end{bmatrix} \succeq 0,$$

and $S = Q + A^T SA - X.$

A popular alternative is to adapt the value iteration algorithm to this setting by iteratively applying the right side of the aforementioned equation as an operator. Concretely, starting from $S_0 \succeq 0$, we choose

$$S_{t+1} = Q + A^T S_t A - A^T S_t B \left(R + B^T S_t B \right)^{-1} B^T S_t A.$$

5.4 Robust Control

The solution to the LQR problem is a hallmark of optimal control theory and one of the most widely used techniques in control. However, the assumption that the disturbances are stochastic and i.i.d. is a strong one, and numerous attempts have been made over the years to generalize the noise model.

One of the most important and prominent generalizations is that of robust control, also known as H_∞-control. The goal of H_∞-control is to come up with a more robust solution by considering the optimal solution vs. any possible perturbation sequence from a certain family. Formally, our aim is to solve the following:

$$\min_{\mathbf{u}_{1:T}(\mathbf{x})} \max_{\mathbf{w}_{1:T}} \sum_{t=1}^{T} \mathbf{x}_t^\top Q \mathbf{x}_t + \mathbf{u}_t^\top R \mathbf{u}_t$$
$$\text{s.t. } \mathbf{x}_{t+1} = A_t \mathbf{x}_t + B_t \mathbf{u}_t + \mathbf{w}_t,$$

where we use the exact same notation as in previous sections.

Notice that this problem does not fall into the category of a Markov decision process, since the $\mathbf{w}_t$'s are no longer stochastic. Therefore, the typical Bellman optimality criterion does not apply.

Instead, the appropriate techniques to address this problem come from the theory of dynamical games. The above is, in fact, a two-player zero sum game. The first player tries to minimize the costs, and the other attempts to maximize the costs, both players subject to the dynamics evolution constraint.

Using techniques from dynamic games, the following can be shown. References are provided in the bibliographic section.

Theorem 5.4 *For the H_∞ control problem with quadratic convex loss functions, the optimal solution is a linear policy of the state given by*

$$\mathbf{u}_t = K_t \mathbf{x}_t.$$

The proof is somewhat more involved than the LQR derivation and is omitted from this text.

5.5 Bibliographic Remarks

Optimal control is considered to be the hallmark of control theory, and numerous excellent introductory texts cover it in detail, for example, Hespanha (2018) and Stengel (1994).

The LQR problem and its solution were introduced in the seminal work of Kalman (1960), who also introduced the eponymous filtering procedure for linear estimation.

While seemingly a natural corollary of equivalent results for MDPs, the existence of value functions for linear control is made subtle by the continuous nature of the state space and stabilizability, or the lack thereof. A discussion of sufficient and necessary conditions for the existence of solutions to Riccati equations can be found in Bertsekas (2007) and Hassibi et al. (1996).

The development of H_∞ control was driven by an observation made in the seminal paper of Doyle (1978) that, unlike in the fully observed case, the solution to the optimal control problem for partially observed linear systems (which we study in later chapters) is not naturally robust to misspecification of system parameters. This suggests that robustness specifications constitute yet another axis of performance for control algorithms and may at times be at odds with the objective of optimal control. For an in-depth treatment of H_∞ control, see Zhou et al. (1996b). The text Başar and Bernhard (2008) presents a succinct derivation of robust controllers entirely in the state space, as the solution to a dynamic game between the controller and an adversary; see also Tu (2018) for a recent exposition.

Although developed in the linear setting, adaptations of the LQR controller are also widely used to control nonlinear systems (Levine and Koltun, 2013; Li, 2005).

5.6 Exercises

5.1. Prove Theorem 5.2 for a time-varying linear dynamical system.

5.2. Derive the solution to the infinite-horizon LQR with zero-mean Gaussian noise.

5.3. Extend the proof of Theorem 5.2 to positive semidefinite Q, R, rather than positive definite matrices.

5.4. Prove Theorem 5.4 as a dynamic two-player zero-sum game, where a unique saddle-point solution exists. Hint: you can formulate this problem as a min-max optimization problem within a finite horizon.

5.5. Prove that the function $f(\mathbf{x}) = \mathbf{x}^\top M \mathbf{x}$ is convex if and only if $M \succeq 0$ is positive semidefinite.

PART II

BASICS OF ONLINE CONTROL

6
Regret in Control

In the first part of this book and in Chapter 5 we studied linear dynamical systems with quadratic costs, namely the LQR problem. We saw that in this case the optimal control takes a very special form: It is a linear function of the state. The Markovian property of the Markov decision process framework for reinforcement learning suggests that the optimal policy is always a function of the current state. It is characterized by the Bellman optimality equation and can be computed in complexity proportional to the number of states and actions using, for example, the value iteration algorithm, which we analyzed in a previous chapter.

However, this changes when we deviate from the MDP framework, which may be warranted for a variety of reasons:

(i) The state is not fully observable.
(ii) The reward functions change online.
(iii) The transition mapping is not fixed and changes online.
(iv) The transitions or costs are subject to adversarial noise.

In such scenarios involving significant deviations from the MDP framework, classical notions of optimality may be illdefined. Or if they do exist, the optimal policy may no longer be a mapping from immediate state to action. Rather, the learner may be required to use other signals available to them when choosing an action that is most beneficial in the long run. To this we add a significant consideration: The computation of optimal **state-based policies may be computationally intractable**.

This motivates the current chapter, which considers regret analysis as a computationally tractable alternative to optimal or robust control.

We start by considering regret in the decision-making framework of online convex optimization. We then consider several classes of control policies and the relationships between them in the context of regret analysis.

6.1 Online Convex Optimization

In this section, we give a brief introduction to the theory of online convex optimization. The reader is referred to the bibliographic section for a comprehensive survey.

In the decision-making framework of online convex optimization, a player iteratively chooses a point in a convex set $\mathcal{K} \subseteq \mathbb{R}^n$ and suffers a loss according to an adversarially chosen loss function $f_t : \mathcal{K} \mapsto \mathbb{R}$.

Let $\mathcal{A}$ be an algorithm for OCO, which maps a certain game history to a decision in the decision set

$$\mathbf{x}_t^{\mathcal{A}} = \mathcal{A}(f_1, \ldots, f_{t-1}) \in \mathcal{K}.$$

We define the regret of $\mathcal{A}$ after T iterations as

$$\operatorname{regret}_T(\mathcal{A}) = \sup_{\{f_1,\ldots,f_T\} \subseteq \mathcal{F}} \left\{ \sum_{t=1}^T f_t(\mathbf{x}_t^{\mathcal{A}}) - \min_{\mathbf{x} \in \mathcal{K}} \sum_{t=1}^T f_t(\mathbf{x}) \right\}. \quad (6.1)$$

If the algorithm is clear from the context, we henceforth omit the superscript and denote the algorithm's decision at time t simply as $\mathbf{x}_t$. Intuitively, an algorithm performs well if its regret is sublinear as a function of T (i.e., $\operatorname{regret}_T(\mathcal{A}) = o(T)$), since this implies that, on average, the algorithm performs as well as the best fixed strategy in hindsight. We then describe a basic algorithm for online convex optimization.

6.1.1 Online Gradient Descent

Perhaps the simplest algorithm that applies to the most general setting of online convex optimization is online gradient descent (OGD). The pseudocode is given in Algorithm 6.1.

Algorithm 6.1 Online Gradient Descent

1: Input: Convex set $\mathcal{K}$, T, $\mathbf{x}_1 \in \mathcal{K}$, step sizes $\{\eta_t\}$
2: **for** $t = 1$ to T **do**
3:     Play $\mathbf{x}_t$ and observe cost $f_t(\mathbf{x}_t)$.
4:     Update and project:

$$\mathbf{y}_{t+1} = \mathbf{x}_t - \eta_t \nabla f_t(\mathbf{x}_t)$$
$$\mathbf{x}_{t+1} = \arg\min_{\mathbf{x} \in \mathcal{K}} \|\mathbf{x} - \mathbf{y}_{t+1}\|^2 = \prod_{\mathcal{K}} [\mathbf{y}_{t+1}]$$

5: **end for**

6.1 Online Convex Optimization

In each iteration, the algorithm takes a step from the previous point in the direction of the gradient of the previous cost. This step may result in a point outside the underlying convex set. In such cases, the algorithm projects the point back to the convex set, that is, finds its closest point in the convex set. Despite the fact that the next cost function may be completely different from the costs observed thus far, the regret attained by the algorithm is sublinear. This is formalized in the following theorem. Here we define the constants D and G as follows:

- D = Euclidean diameter of the decision set, that is,

$$\max_{\mathbf{x},\mathbf{y}\in\mathcal{K}} \|\mathbf{x}-\mathbf{y}\|.$$

- G = an upper bound on the Lipschitz constant of the cost functions f_t, or equivalently, the subgradient norms of these functions. Mathematically,

$$\forall t, \ \|\nabla f_t(\mathbf{x}_t)\| \leq G.$$

Theorem 6.1 *Online gradient descent with step sizes* $\{\eta_t = \frac{D}{G\sqrt{t}}, \ t \in [T]\}$ *guarantees the following for all* $T \geq 1$:

$$\text{regret}_T = \sum_{t=1}^{T} f_t(\mathbf{x}_t) - \min_{\mathbf{x}^\star \in \mathcal{K}} \sum_{t=1}^{T} f_t(\mathbf{x}^\star) \leq \frac{3}{2} GD\sqrt{T}.$$

Proof Let $\mathbf{x}^\star \in \arg\min_{\mathbf{x}\in\mathcal{K}} \sum_{t=1}^{T} f_t(\mathbf{x})$. Define $\nabla_t \stackrel{\text{def}}{=} \nabla f_t(\mathbf{x}_t)$. By convexity

$$f_t(\mathbf{x}_t) - f_t(\mathbf{x}^\star) \leq \nabla_t^\top (\mathbf{x}_t - \mathbf{x}^\star). \tag{6.2}$$

We first upper-bound $\nabla_t^\top (\mathbf{x}_t - \mathbf{x}^\star)$ using the update rule for $\mathbf{x}_{t+1}$ and by the Pythagorean theorem, we have the projections only decrease the distance to the set $\mathcal{K}$:

$$\|\mathbf{y}_{t+1}-\mathbf{x}^\star\|^2 = \left\|\prod_{\mathcal{K}}(\mathbf{x}_t - \eta_t \nabla_t) - \mathbf{x}^\star\right\|^2 \leq \|\mathbf{x}_t - \eta_t \nabla_t - \mathbf{x}^\star\|^2. \tag{6.3}$$

Hence,

$$\|\mathbf{y}_{t+1}-\mathbf{x}^\star\|^2 \leq \|\mathbf{x}_t - \mathbf{x}^\star\|^2 + \eta_t^2 \|\nabla_t\|^2 - 2\eta_t \nabla_t^\top (\mathbf{x}_t - \mathbf{x}^\star),$$

$$2\nabla_t^\top (\mathbf{x}_t - \mathbf{x}^\star) \leq \frac{\|\mathbf{x}_t - \mathbf{x}^\star\|^2 - \|\mathbf{x}_{t+1} - \mathbf{x}^\star\|^2}{\eta_t} + \eta_t G^2. \tag{6.4}$$

Summing (6.2) and (6.4) from $t = 1$ to T, and setting $\eta_t = \frac{D}{G\sqrt{t}}$ (with $\frac{1}{\eta_0} \stackrel{\text{def}}{=} 0$):

$$2\left(\sum_{t=1}^{T} f_t(\mathbf{x}_t) - f_t(\mathbf{x}^\star)\right) \leq 2\sum_{t=1}^{T} \nabla_t^\top (\mathbf{x}_t - \mathbf{x}^\star)$$

$$\leq \sum_{t=1}^{T} \frac{\|\mathbf{x}_t - \mathbf{x}^\star\|^2 - \|\mathbf{x}_{t+1} - \mathbf{x}^\star\|^2}{\eta_t} + G^2 \sum_{t=1}^{T} \eta_t$$

$$\leq \sum_{t=1}^{T} \|\mathbf{x}_t - \mathbf{x}^\star\|^2 \left(\frac{1}{\eta_t} - \frac{1}{\eta_{t-1}}\right) + G^2 \sum_{t=1}^{T} \eta_t \qquad \frac{1}{\eta_0} \stackrel{\text{def}}{=} 0,$$

$$\|\mathbf{x}_{T+1} - \mathbf{x}^*\|^2 \geq 0$$

$$\leq D^2 \sum_{t=1}^{T} \left(\frac{1}{\eta_t} - \frac{1}{\eta_{t-1}}\right) + G^2 \sum_{t=1}^{T} \eta_t$$

$$\leq D^2 \frac{1}{\eta_T} + G^2 \sum_{t=1}^{T} \eta_t \qquad \text{telescoping series}$$

$$\leq 3DG\sqrt{T}.$$

The last inequality follows since $\eta_t = \frac{D}{G\sqrt{t}}$ and $\sum_{t=1}^{T} \frac{1}{\sqrt{t}} \leq 2\sqrt{T}$. □

The online gradient descent algorithm is straightforward to implement, and updates take linear time given the gradient. However, in general, the projection step may take significantly longer via convex optimization.

6.2 Regret for Control

The most important consideration to explore beyond classical models of control, such as optimal and robust control, is computational. Even in linear control with convex loss functions, the optimal policy can be complex to describe – see the exercises section.

A prominent approach to deal with computational difficulty is to consider a less stringent criterion for optimality. Instead of the optimal policy, we can consider all policies in a certain *policy class* and try to compete with the optimal policy within the class in terms of regret.

As alluded to earlier, the most natural classes of policies are computationally intractable. We thus resort to learning new classes that are computationally favorable. But to relate the performance of our methods, we require a language of comparing different policy classes. It is crucial to be able to relate

6.2 Regret for Control

the representation power of different policy classes. This motivates our main comparison metric as follows.

Definition 6.2 We say that a class of policies Π_1 ε-approximates class Π_2 if the following is satisfied: For every $\pi_2 \in \Pi_2$, there exists $\pi_1 \in \Pi_1$ such that for all sequences of T disturbances and cost functions, it holds that

$$\sum_{t=1}^{T} \left| c_t(\mathbf{x}_t^{\pi_1}, \mathbf{u}_t^{\pi_1}) - c_t(\mathbf{x}_t^{\pi_2}, \mathbf{u}_t^{\pi_2}) \right| \le T\varepsilon.$$

The significance of this definition is that if a simple policy class approximates another more complicated class, then it suffices to consider only the first. The usefulness of this definition comes from its implications for regret minimization, according to Definition 1.3, as follows.

Lemma 6.3 *Let policy class Π_1 ε-approximate policy class Π_2. Then for any control algorithm $\mathcal{A}$ we have*

$$\mathrm{regret}_T(\mathcal{A}, \Pi_2) \le \mathrm{regret}_T(\mathcal{A}, \Pi_1) + \varepsilon T.$$

Proof Recall that for policy class Π and algorithm $\mathcal{A}$,

$$\mathrm{regret}_T(\mathcal{A}, \Pi) = \max_{\mathbf{w}_{1:T} : \|\mathbf{w}_t\| \le 1} \left(\sum_{t=1}^{T} c_t(\mathbf{x}_t, \mathbf{u}_t) - \min_{\pi \in \Pi} \sum_{t=1}^{T} c_t(\mathbf{x}_t^\pi, \mathbf{u}_t^\pi) \right).$$

It follows from Definition 6.2 that

$$\min_{\pi_2 \in \Pi_2} \sum_{t=1}^{T} c_t(\mathbf{x}_t^{\pi_2}, \mathbf{u}_t^{\pi_2}) = \sum_{t=1}^{T} c_t(\mathbf{x}_t^{\pi_2^*}, \mathbf{u}_t^{\pi_2^*}) \qquad \pi_2^* = \arg\min_{\pi \in \Pi_2} \sum_t c_t(\mathbf{x}_t^\pi, \mathbf{u}_t^\pi)$$

$$\ge \sum_{t=1}^{T} c_t(\mathbf{x}_t^{\pi_1}, \mathbf{u}_t^{\pi_1}) - \varepsilon T \qquad \text{Definition 6.2}$$

$$\ge \min_{\pi_1 \in \Pi_1} \sum_{t=1}^{T} c_t(\mathbf{x}_t^{\pi_1}, \mathbf{u}_t^{\pi_1}) - \varepsilon T.$$

Thus, for any fixed disturbance sequence $\mathbf{w}_{1:T}$,

$$\sum_{t=1}^{T} c_t(\mathbf{x}_t, \mathbf{u}_t) - \min_{\pi_2 \in \Pi_2} \sum_{t=1}^{T} c_t(\mathbf{x}_t^{\pi_2}, \mathbf{u}_t^{\pi_2})$$

$$\le \sum_{t=1}^{T} c_t(\mathbf{x}_t, \mathbf{u}_t) - \left(\min_{\pi_1 \in \Pi_1} \sum_{t=1}^{T} c_t(\mathbf{x}_t^{\pi_1}, \mathbf{u}_t^{\pi_1}) - \varepsilon T \right)$$

$$= \left(\sum_{t=1}^{T} c_t(\mathbf{x}_t, \mathbf{u}_t) - \min_{\pi_1 \in \Pi_1} \sum_{t=1}^{T} c_t(\mathbf{x}_t^{\pi_1}, \mathbf{u}_t^{\pi_1}) \right) + \varepsilon T.$$

Since this holds for any fixed disturbance sequence $\mathbf{w}_{1:T}$, it also holds for the maximum over all disturbance sequences $\mathbf{w}_{1:T}$ with $\|\mathbf{w}_t\| \leq 1$. □

We now proceed to define and study approximation relationships between different policy classes.

6.3 Expressivity of Control Policy Classes

In this section, we formally relate the power of different complexity classes for linear time-invariant linear dynamical systems. The definitions and results can be extended to time-varying linear dynamical systems, and recall from Chapter 4 that time-varying linear dynamical systems were proposed by Lyapunov as a general methodology to study nonlinear dynamics.

For the rest of this section we assume full observation of the state and let

$$\mathbf{x}_{t+1} = A\mathbf{x}_t + B\mathbf{u}_t + \mathbf{w}_t.$$

A sequence of stabilizing linear controllers for such a system is denoted by $\{K_t\}$. Schematically, the relationships we describe below are given in Figure 6.1 for fully observed systems. In later chapters of this text, we study policies for partially observed systems.

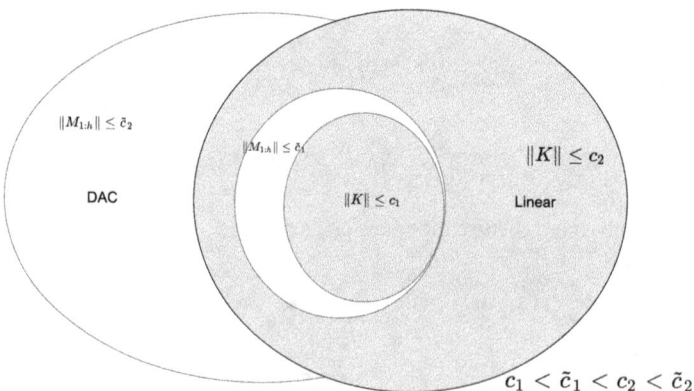

Figure 6.1 Schematic relationship between linear disturbance-action control policies and linear state-action policies of increasing diameter for linear time invariant systems.

6.3 Expressivity of Control Policy Classes

6.3.1 Linear Policies

The most basic and well-studied policy classes for control of linear dynamical systems are linear policies, that is, those that choose a control signal as a (fixed) linear function of the state. The most important subcases of linear policies are those that are γ-stabilizing, meaning that the signal they create does not cause an unbounded state. This was formalized in Definition 2.4. Formally,

Definition 6.4 (Linear policies) A linear controller $K \in \mathbb{R}^{d_x \times d_u}$ is a linear operator that maps the state to the control. We denote by Π_γ^L the set of all linear controllers that are γ-stabilizing according to Definition 2.4 and are bounded in Frobenius norm by $\|K\| \leq \gamma$.

6.3.2 Disturbance Action Controllers

We next define the most important set of policies in online nonstochastic control, known as disturbance action controllers, or DACs. The main advantages of DACs, as opposed to linear controllers, are that the cost of control is convex in their parameterization and thus allows for provably efficient optimization. Moreover, this set of policies is stronger than linear policies in the sense that it ε-approximates them, as we show in the following.

Definition 6.5 (Disturbance action controller) A (linear) disturbance action policy $\pi_{K_{1:T}, M_{1:h}}$ is parameterized by the matrices $M_{0:h} = [M_0, M_1, \ldots, M_h]$ and stabilizing controllers $\{K_t\}$. It outputs control $\mathbf{u}_t^\pi$ at time t according to the rule

$$\mathbf{u}_t^\pi = K_t \mathbf{x}_t + \sum_{i=1}^{h} M_i \mathbf{w}_{t-i} + M_0.$$

Without loss of generality, we take $M_0 = 0$, as we can add a fictitious unit coordinate to the perturbations $\{\mathbf{w}_t\}$, and modify the system accordingly. This will allow for a constant shift in the control.

Denote by $\Pi_{h,\gamma}^D$ the set of all disturbance action policies with $K_{1:T}$ being γ-stabilizing linear controllers for the system and

$$\sum_{i=1}^{h} \|M_i\| \leq \gamma.$$

The class of DACs always produces a stabilizing control signal whenever $K_{1:T}$ are sequentially stabilizing for the entire system $\{A_t, B_t\}$. This important fact can be seen in the following lemma.

Lemma 6.6 *Any DAC policy in $\Pi_{h,\gamma}^D$ is γ^2-stabilizing.*

Proof For convenience, denote by $M_0 = I$, and the fictitious noise term $\tilde{\mathbf{w}}_t = \sum_{i=0}^{h} M_i \mathbf{w}_{t-i}$. First, due to the bound on $M_{1:h}$ and $\mathbf{w}_t$, we have that

$$\|\tilde{\mathbf{w}}_t\| = \left\|\sum_{i=0}^{h} M_i \mathbf{w}_{t-i}\right\| \le \sum_{i=0}^{h} \|M_i\| \|\mathbf{w}_{t-i}\| \le \gamma,$$

for $\|M\|$ being the spectral norm of M.

Next, since the K_t's are stabilizing, we can bound the historically distant terms as follows. For simplicity, assume that $K_t = K$ are all equal. Observe that the state at time t, following any DAC policy π, is given by

$$\begin{aligned} \mathbf{x}_{t+1} &= A\mathbf{x}_t + B\mathbf{u}_t + \mathbf{w}_t \\ &= A\mathbf{x}_t + B\left(K\mathbf{x}_t + \sum_{i=1}^{h} M_i \mathbf{w}_{t-i}\right) + \mathbf{w}_t \\ &= (A + BK)\mathbf{x}_t + B\sum_{i=0}^{h} M_i \mathbf{w}_{t-i} \qquad \text{using } M_0 = I \\ &= \sum_{j=0}^{t} \left((A + BK)^j B \sum_{i=0}^{h} M_i \mathbf{w}_{t-i-j}\right) \\ &= \sum_{j=0}^{t-1} \left((A + BK)^j B \tilde{\mathbf{w}}_{t-j}\right). \end{aligned}$$

We can thus view the states x_t as generated by a γ-stabilizing policy over noises that are of magnitude γ, and hence bounded by γ^2. □

The above lemma is already a good indication about DACs: They generate controls that at least do not cause a blow-up. But how expressive are these policies in terms of reaching the optimal solution?

One of the main appeals of the DAC class is the following approximation lemma, which shows that for time-invariant systems DACs are in fact a stronger class than linear controllers. Notice that the K_t terms for the DAC are only assumed to be stabilizing and not necessarily optimal for the actual sequence.

Lemma 6.7 *Consider a time-invariant linear dynamical system $\{A, B\}$. The class $\Pi_{h,\gamma}^D$ ε-approximates the class $\Pi_{\gamma'}^L$ for $h = \Omega(\gamma' \log(\frac{\gamma'}{\varepsilon}))$ and $\gamma = \gamma'^2$.*

Proof Disturbance–action policies use a γ-stabilizing linear policy on which they superimpose linear functions of disturbances – fix such a policy, and let K_0 be this linear stabilizing component. The state sequence thus produced is identical to that obtained by running a DAC with $\mathbf{0}$ as the stabilizing controller on the system $(A + BK_0, B)$. Similarly, the state sequence produced by the

6.3 Expressivity of Control Policy Classes

execution of a linear policy K on (A, B) is identical to that of the linear policy $K - K_0$ on the system $(A + BK_0, B)$.

Lemma 6.8 *Let $\mathbf{x}_t^\pi(A, B), \mathbf{u}_t^\pi(A, B)$ be the state–action sequence produced by a policy π when executed on a linear dynamical system (A, B). For any $\varepsilon > 0$, if two policies π_1, π_2 satisfy for all t that*

$$\|\mathbf{x}_t^{\pi_1}(A + BK_0, B) - \mathbf{x}_t^{\pi_2}(A + BK_0, B)\| \le \varepsilon,$$

$$\|\mathbf{u}_t^{\pi_1}(A + BK_0, B) - \mathbf{u}_t^{\pi_2}(A + BK_0, B)\| \le \varepsilon$$

when executed on some dynamical system $(A + BK_0, B)$, then when the corresponding policies π_1', π_2', that is, π_1 and π_2 with K_0 superimposed, are executed on the dynamical system (A, B), the iterates guarantee

$$\max\{\|\mathbf{x}_t^{\pi_1'}(A, B) - \mathbf{x}_t^{\pi_2'}(A, B)\|,$$
$$\|\mathbf{u}_t^{\pi_1'}(A, B) - \mathbf{u}_t^{\pi_2'}(A, B)\|\} \le (1 + \|K_0\|)\varepsilon.$$

With the aid of the above lemma, it will suffice to guarantee that the state–action sequence produced by the linear policy K and an appropriately chosen DAC with the stabilizing part set to $\mathbf{0}$ on an intrinsically γ-stable system (A, B) are similar enough. To prove this, we map any linear policy K to a disturbance–action policy that approximates it arbitrarily well for some history parameter h. Specifically, for a given system A, B and linear policy K, let π_K be the policy given by

$$\forall i \in [0, h], \ M_i = K(A + BK)^i.$$

Now consider any trajectory that results from applying the linear policy for the given system. Let $\mathbf{x}_{1:T}, \mathbf{u}_{1:T}$ be the states and controls of the linear policy K for a certain trajectory, as affected by a given noise signal. Let $\tilde{\mathbf{x}}_{1:T}, \tilde{\mathbf{u}}_{1:T}$ be the set of states and controls that would have been generated with the same noise sequence, but with the policy π_K. We first observe that control sequences produced by K and π_K are exponentially close.

$$\mathbf{u}_{t+1} = K\mathbf{x}_{t+1} = K(A\mathbf{x}_t + B\mathbf{u}_t + \mathbf{w}_t)$$
$$= K(A + BK)\mathbf{x}_t + K\mathbf{w}_t$$
$$= \sum_{i=0}^{t} K(A + BK)^i \mathbf{w}_{t-i}.$$

Now, we define $Z_t = \sum_{i=h+1}^{t} K(A + BK)^i \mathbf{w}_{t-i}$, and continue as shown.

$$\mathbf{u}_{t+1} = \sum_{i=0}^{h} K(A+BK)^i \mathbf{w}_{t-i} + Z_t$$

$$= \sum_{i=0}^{h} M_i \mathbf{w}_{t-i} + Z_t$$

$$= \tilde{\mathbf{u}}_{t+1} + Z_t.$$

The magnitude of this residual term Z_t can be bounded, due to the existence of $\frac{\kappa}{\delta} \leq \gamma'$ from Exercise 4.5, as

$$|Z_t| \leq \left\| \sum_{i=h+1}^{t} K(A+BK)^i \mathbf{w}_{t-i} \right\|$$

$$\leq \sum_{i=h+1}^{t} \kappa(1-\delta)^{i+1} \|K\| \|\mathbf{w}_{t-i}\|$$

$$\leq \kappa^2 \sum_{i=h+1}^{\infty} (1-\delta)^i \leq \kappa^2 \int_{i=h}^{\infty} e^{-\delta i} di$$

$$= \kappa^2 \frac{1}{\delta} e^{-\delta h} \leq \varepsilon.$$

We thus conclude that $\|\mathbf{u}_{t+1} - \tilde{\mathbf{u}}_{t+1}\| \leq \varepsilon$.

Similarly, for the corresponding state sequences, first note that

$$\mathbf{x}_{t+1} = \sum_{i=0}^{t} A^i (\mathbf{w}_{t-i} + B\mathbf{u}_{t-i}).$$

Therefore, since $\mathbf{x}_t$ and $\tilde{\mathbf{x}}_t$ share the same perturbation sequences as a part of their description, their difference can be bounded as a geometrically weighted sum of the difference in control inputs $\mathbf{u}_t$ and $\tilde{\mathbf{u}}_t$ that constitute them. Concretely, we have

$$\|\mathbf{x}_t - \tilde{\mathbf{x}}_t\| = \left\| \sum_{i=1}^{t} A^{i-1} B(\mathbf{u}_{t-i} - \tilde{\mathbf{u}}_{t-i}) \right\|$$

$$\leq \left\| \sum_{i=0}^{t} A^i B \right\| \max_{i \leq t} \|\mathbf{u}_t - \tilde{\mathbf{u}}_i\| \leq \frac{\kappa}{\delta} \varepsilon,$$

by using the γ-stability of (A, B). The approximation in terms of states and controls implies ε-approximation of policy class for any Lipschitz cost sequence, concluding the proof. $\square$

6.4 Bibliographic Remarks

Regret minimization, online learning, and online convex optimization have a rich history and have been studied for decades in the machine learning and game theory literature. The reader is referred to the comprehensive texts of Cesa-Bianchi and Lugosi (2006) and Hazan (2016) for a detailed treatment.

Linear state-feedback policies are a mainstay of control theory in both applications, due to their simplicity, and in theory, due to their optimality in deterministic and stochastic settings, as demonstrated in Kalman (1960). The methods for computing such policies are often based on dynamic programming. Often motivated by the need to impose constraints beyond optimality, such as robustness, various other policy parameterizations were introduced. An early example is the Youla parameterization, sometimes called Q-parameterization (Youla et al., 1976). Later, the introduction of system-level synthesis (SLS) (Anderson et al., 2019; Doyle et al., 2017), in a similar vein, proved useful for distributed control, especially given its emphasis on state-space representations. For a comparison of these and additional ones such as input-output parameterizations (Furieri et al., 2019), see Zhang et al. (2023) and Zheng et al. (2020).

The disturbance action control parameterization, introduced in Agarwal et al. (2019c), is similar to system-level synthesis. Like SLS, it is a convex parameterization that is amenable to the tools of convex optimization. However, unlike SLS, it is a more compact representation that has logarithmically many parameters in its natural representation in terms of the desired approximation guarantee against linear policies. This makes it better suited for machine learning.

The study of regret in the context of control has roots in the work of Abbasi-Yadkori and Szepesvári (2011) and Abbasi-Yadkori et al. (2014). These works, as well as many that followed (Agarwal et al., 2019b; Cohen et al., 2018; Dean et al., 2018; Mania et al., 2019; Simchowitz and Foster, 2020), study regret in a stochastic setting, where the perturbations are stochastic. In such settings, optimal control is well defined. The full power of regret analysis for control, involving adversarial perturbations, adversarial loss functions, and computational considerations, was initiated in Agarwal et al. (2019c). The notion of regret as applied to policy classes was implicit in the work of Even-Dar et al. (2004). The regret metric when applied to policies is different from the counterfactual notion of policy regret defined by Arora et al. (2012).

6.5 Exercises

6.1. Let us say that the total number of time steps T is known in advance. Consider an implementation of online gradient descent where the step size is chosen as $\eta_t = D/G\sqrt{T}$ and therefore is uniform over time. Prove that the resultant regret is bounded by $GD\sqrt{T}$, which improves upon Theorem 6.1 by a constant factor. In fact, for ease of exposition, we will use this uniform step size scheme going forward.

6.2. In this exercise, we prove that the optimal solution to a linear control problem can be complex to describe.

(i) Consider a linear dynamical system in one dimension and loss functions of the form $f_t(\mathbf{x}) = \max\{0, \mathbf{x} - c\}$, for $c_t \in \mathbb{R}$. Show that the optimal policy in hindsight can have as many domains as loss functions.
(ii) Consider a similar setting in higher dimensions and give a bound on the number of domains as a function of the number of loss functions and the dimension.

6.3. Consider the setting of time-varying linear dynamical systems, that is,

$$\mathbf{x}_{t+1} = A_t \mathbf{x}_t + B_t \mathbf{u}_t + \mathbf{w}_t.$$

Prove that any DAC according to Definition 6.5 is stable, that is, prove the analogue of Lemma 6.6 for time-varying systems.

6.4. Prove Lemma 6.8.

7
Online Nonstochastic Control

In this chapter, we explore recent advances in machine learning that have significantly reshaped the control problem. These developments mark a major departure from classical control theory and, in many cases, have led to methods that not only compete with, but often outperform traditional approaches. The key characteristics of these modern techniques are as follows:

(i) **Online Learning Over Precomputed Policies** Unlike classical control, which relies on precomputing an optimal policy, online control adopts adaptive learning techniques to update policies dynamically. This aligns with the broader framework of adaptive control, where policies evolve in response to observed behavior of the system.

(ii) **Regret Minimization as a Performance Metric** While adaptive control focuses on learning a stabilizing policy, online control introduces a rigorous performance metric: regret, defined as the difference in cost between the learned policy and the best policy in hindsight. Unlike classical approaches, online nonstochastic control algorithms provide finite-time regret guarantees, ensuring competitive performance even under adversarial conditions.

(iii) **General Loss Functions and Adversarial Perturbations** Classical control typically assumes quadratic cost functions with known structures. In contrast, online nonstochastic control allows for general convex loss functions, which may be selected adversarially. This greater flexibility enables controllers to adapt to more diverse and complex real-world scenarios.

(iv) **Improper Learning via Convex Relaxation** A fundamental breakthrough in online nonstochastic control is the use of improper learning, where the algorithm competes against a given policy class while learning from a broader class of policies. Using convex relaxation techniques, these methods bypass computational hardness barriers that traditionally limit optimal control, and they are accompanied by efficient algorithms with strong theoretical guarantees.

These innovations redefine how we approach control problems, shifting the focus from static policy optimization to adaptive, regret-minimizing strategies that operate effectively in dynamic and adversarial environments.

7.1 From Optimal and Robust to Online Control

Classical control theory has been mainly shaped by two paradigms: **optimal control** and **robust control**. These approaches have been instrumental in various engineering applications, but they rely on specific assumptions about disturbances and system dynamics.

- **Optimal control** assumes that disturbances follow a known stochastic model, often Gaussian. The objective is to design a control policy that minimizes the expected cumulative cost. Notable methods such as the linear quadratic regulator (LQR) and stochastic dynamic programming fall into this category.
- **Robust control**, in contrast, takes a worst-case perspective. It ensures stability and performance even under the most adverse conditions, assuming perturbations belong to a predefined uncertainty set. A key example is H-infinity control.

Although these frameworks have been successful, real-world control tasks often involve uncertainties that do not fit neatly into either a probabilistic model or a worst-case adversarial setting. Consider the challenge of controlling a drone in an unpredictable environment:

- A stochastic model of wind disturbances may be unrealistic since actual wind patterns can exhibit long-term correlations or abrupt shifts.
- A worst-case robust approach may result in excessively conservative control, leading to inefficient and overly cautious maneuvers.

These limitations motivate the need for a more adaptive approach – one that can learn and adjust dynamically based on observed disturbances rather than relying on rigid assumptions. This leads us to **online nonstochastic control**, a fundamentally different paradigm where controllers adapt in real-time without assuming a fixed model for disturbances.

7.1.1 A Shift to Online Nonstochastic Control

Rather than relying on predefined stochastic or worst-case assumptions, online nonstochastic control introduces a new approach based on sequential decision making and regret minimization. The core principles are the following.

7.2 The Online Nonstochastic Control Problem

(i) **No Assumption on the Noise Model** Disturbances (e.g., wind forces, external perturbations) are treated as an arbitrary sequence, potentially chosen adversarially. The controller must adapt dynamically rather than relying on a fixed distribution.

(ii) **Learning and Adaptation Over Time** Instead of precomputing an optimal policy, online control algorithms iteratively refine their strategies using techniques from online convex optimization and sequential decision making.

(iii) **Regret Minimization as the Performance Metric** Unlike optimal control, which minimizes expected cost, or robust control, which prepares for the worst case, online control seeks to minimize regret. Regret quantifies how much worse the controller performs compared to the best policy in hindsight.

To build intuition, consider the problem of controlling an autonomous vehicle in an unknown city. If the vehicle had complete foresight of future traffic conditions, it could calculate the optimal route in advance. However, without such knowledge, the vehicle must continuously adjust its route based on real-time observations. Regret measures how far the vehicle's performance deviates from the best possible path that could have been chosen in hindsight.

This shift in perspective enables online controllers to handle a wide range of real-world disturbances, achieving adaptability, robustness, and computational efficiency without requiring strong assumptions about uncertainty.

In the following sections, we formally define the online nonstochastic control problem and introduce key algorithms that provide provable performance guarantees.

7.2 The Online Nonstochastic Control Problem

The shift from classical control to online nonstochastic control requires a fundamental rethinking of how we handle uncertainty. Unlike optimal control, which assumes disturbances follow a known probabilistic model, or robust control, which plans for the worst case, online nonstochastic control operates in a more general and challenging setting.

In this framework, both disturbances and cost functions are chosen by an adversary, meaning that they may follow no statistically reliable pattern and could even be designed to hinder the controller's decisions at every step. As a result, the goal is no longer to compute a fixed optimal control policy but rather to learn and adapt dynamically in response to an evolving sequence of perturbations and costs.

A fundamental difficulty in control is the computational tractability of finding the best policy. Classical control methods often rely on dynamic programming, which, in general, suffers from the curse of dimensionality. Even in simple settings, computing an optimal policy can be intractable due to the complexity of solving constrained optimization problems on trajectories. As pointed out by Rockafellar, finding an optimal control policy in arbitrary convex cost settings can be computationally prohibitive, particularly when state and control constraints are involved.

To address this, online nonstochastic control shifts the focus from computing an optimal policy to competing with a benchmark class of policies using regret minimization. Instead of optimizing a predefined objective function, the controller seeks to minimize its total loss relative to the best policy that could have been chosen in hindsight. This approach bypasses the need to solve an intractable dynamic optimization problem in advance while still achieving performance guarantees over time.

The fundamental challenge in online nonstochastic control can be summarized as follows:

- The controller does not have prior knowledge of the cost functions or disturbances.
- At each time step, the controller must select a control input based only on past observations.
- The performance of the controller is evaluated against a benchmark: the best fixed policy that could have been chosen in hindsight.

This formulation naturally connects to online learning and online convex optimization, where decisions must be made sequentially in uncertain environments. We now provide a formal definition of the problem.

Definition 7.1 Consider a dynamical system evolving according to the equation $\mathbf{x}_{t+1} = f_t(\mathbf{x}_t, \mathbf{u}_t, \mathbf{w}_t)$, where:

- $\mathbf{x}_t \in \mathbb{R}^{d_x}$ is the system state at time t.
- $\mathbf{u}_t \in \mathbb{R}^{d_u}$ is the control input at time t.
- $\mathbf{w}_t \in \mathbb{R}^{d_w}$ represents an external perturbation, which is chosen adversarially.

At each time step, the controller $\mathcal{A}$ selects u_t based on past observations, then incurs a cost defined by an arbitrary convex function $c_t : \mathbb{R}^{d_x} \times \mathbb{R}^{d_u} \to \mathbb{R}$.

The objective is to design a control strategy that minimizes regret, defined as the difference between the total cost incurred by the controller and the best fixed policy in hindsight:

7.3 The Gradient Perturbation Controller

$$\text{regret}_T(\mathcal{A}, \Pi) = \max_{\mathbf{w}_{1:T} : \|\mathbf{w}_t\| \leq 1} \left(\sum_{t=1}^{T} c_t(\mathbf{x}_t, \mathbf{u}_t) - \min_{\pi \in \Pi} \sum_{t=1}^{T} c_t(\mathbf{x}_t^\pi, \mathbf{u}_t^\pi) \right),$$

where Π is a benchmark class of policies, and (x_t^π, u_t^π) represent the state and control sequence obtained under the best policy π in hindsight.

Henceforth, if T, Π, and $\mathcal{A}$ are clear from the context, we refer to regret without quantifiers. The goal of an online control algorithm is to achieve sublinear regret,

$$\text{regret}_T = o(T), \tag{7.1}$$

which ensures that as $T \to \infty$, the controller performs competitively with the best fixed policy.

7.2.1 Improper Learning and Convex Relaxation

A key technique that enables online nonstochastic control algorithms to remain computationally efficient is improper learning through convex relaxation. Instead of directly competing with a predefined class of policies, these algorithms leverage a different policy class that allows for efficient optimization while still achieving strong performance guarantees.

This approach is crucial because enforcing exact constraints on the policy space can lead to computational hardness results, as previously discussed. By relaxing the policy representation, for example, using convex approximations or surrogate loss functions, controllers can sidestep these intractabilities while still competing effectively with the best policy in hindsight.

In the next section, we introduce algorithmic solutions that utilize this principle to achieve low-regret online control under adversarial conditions.

7.3 The Gradient Perturbation Controller

After motivating the framework, defining regret, and exploring the relationships between different policy classes, we are ready to prove the main new algorithm in non-stochastic control.

The algorithm is described in Algorithm 7.1, for time-varying linear dynamical systems. It is a specialization of the template introduced in Algorithm 1.1 from Section 1.6.2. The main idea is to learn the parameterization of a DAC policy using known online convex optimization algorithms, namely online gradient descent. We henceforth prove that this parameterization is convex, allowing us to prove the main regret bound.

The notation we use to describe the algorithm is as follows. We denote by $\mathbf{x}_t(M_{1:h})$ the hypothetical state reached by playing a DAC policy $M_{1:h}$ from the beginning of time. Note that this may not be equal to the true state $\mathbf{x}_t$ of the dynamical system at time t. The counterfactual $\mathbf{u}_t(M_{1:h})$ is defined similarly.

Algorithm 7.1 Gradient Perturbation Controller (GPC)

1: Input: h, η, initialization $M_{1:h}^1 \in \mathcal{K}$.
2: Observe the linear system (A_t, B_t); compute a stabilizing linear controller K_t as per Theorem 4.4.
3: **for** $t = 1 \ldots T$ **do**
4:     Use control $\mathbf{u}_t = K_t \mathbf{x}_t + \sum_{i=1}^{h} M_i^t \mathbf{w}_{t-i}$.
5:     Observe state $\mathbf{x}_{t+1}$; compute noise $\mathbf{w}_t = \mathbf{x}_{t+1} - A_t \mathbf{x}_t - B_t \mathbf{u}_t$.
6:     Record an instantaneous cost $c_t(\mathbf{x}_t, \mathbf{u}_t)$.
7:     Construct loss $\ell_t(M_{1:h}) = c_t(\mathbf{x}_t(M_{1:h}), \mathbf{u}_t(M_{1:h}))$.
      Update $M_{1:h}^{t+1} \leftarrow \prod_{\mathcal{K}} \left[M_{1:h}^t - \eta \nabla \ell_t(M_{1:h}^t) \right]$.
8: **end for**

The GPC algorithm comes with a very strong performance guaranteee: It guarantees vanishing regret vs. the best disturbance action control policy belonging to our reference class in hindsight. According to the relationship between DAC and other policies, this implies vanishing regret versus linear controllers as well. The formal statement is given as follows.

Theorem 7.2 *Assuming that for all $t \in [T]$*

(a) The costs c_t are convex, bounded, and L-Lipschitz.
(b) The matrices $\{A_t, B_t\}$ are γ-stabilizable and have a bounded ℓ_2 norm.

Then the GPC (Algorithm 7.1) ensures that

$$\max_{\mathbf{w}_{1:T}: \|\mathbf{w}_t\| \leq 1} \left(\sum_{t=1}^{T} c_t(\mathbf{x}_t, \mathbf{u}_t) - \min_{\pi \in \Pi^D} \sum_{t=1}^{T} c_t(\mathbf{x}_t^\pi, \mathbf{u}_t^\pi) \right) \leq O(LGD\gamma^2 h \sqrt{T}),$$

where the O notation hides constants that only depend on the norm of A_t, B_t. Furthermore, the time complexity of each loop of the algorithm is linear in the number of system parameters and logarithmic in T.

Note: We henceforth assume that the linear dynamical systems (A_t, B_t) are a linear time-invariant system (A, B) which is γ-stable, and there is no need

7.3 The Gradient Perturbation Controller

for a stabilizing controller $K_t = 0$. The extension to time-varying stabilizable systems is left as Exercise 7.5.

Proof According to Theorem 6.1, the online gradient descent algorithm for convex loss functions ensures that the regret is bounded by $2GD\sqrt{T}$. However, to directly apply the OGD algorithm to our setting, we need to ensure that the following conditions hold true:

- The loss function should be a convex function in the variables $M_{1:h}$.
- We must ensure that optimizing the loss function ℓ_t as a function of parameters $M_{1:h}$ is similar to optimizing the cost c_t as a function of the state and control.

These facts are formalized in the following two lemmas, which we state here and prove at the end of this section.

Lemma 7.3 *The loss functions ℓ_t are convex in the variables $M_{1:h}$.*

Lemma 7.4 *Let G denote a upper bound on the gradients of the L-Lipschitz losses c_t, and let D denote the diameter of the matrices $M_{1:h}$. Then there exists a constant C that depends only on $\|A\|, \|B\|$, such that for all t,*

$$|\ell_t(M_{1:h}^t) - c_t(\mathbf{x}_t, \mathbf{u}_t)| \leq \frac{CL\gamma^2 hD}{\sqrt{T}}.$$

The regret bound of GPC can be derived from these lemmas as follows. Note that G is an upper bound on the gradients of ℓ_t as a function of the parameters $M_{1:h}$, while L is the Lipschitz constant of c_t as a function of the state and control.

$$\sum_{t=1}^{T} c_t(\mathbf{x}_t, \mathbf{u}_t) - \min_{\pi \in \Pi^D} \sum_{t=1}^{T} c_t(\mathbf{x}_t^\pi, \mathbf{u}_t^\pi)$$

$$\leq \sum_{t=1}^{T} \ell_t(M_{1:h}^t) - \min_{\pi \in \Pi^{DAC}} \ell_t(M_{1:h}^\pi) + T \times \frac{L\gamma^2 ChD}{\sqrt{T}} \qquad \text{Lemma 7.4}$$

$$\leq 2GD\sqrt{T} + \sqrt{T}L\gamma^2 ChD \qquad \text{Theorem 6.1}$$

$$\leq O(GDL\gamma^2 h\sqrt{T}).$$

□

7.3.1 Convexity of the Losses

To conclude the analysis, we first show that the loss functions are convex with respect to the variables $M_{1:h}$. This follows since the states and the controls are linear transformations of the variables.

Proof of Lemma 7.3 For simplicity, assume that the initial state $\mathbf{x}_0$ is $\mathbf{0}$. Using $\mathbf{x}_t$ to denote $\mathbf{x}_t(M_{1:h})$ and $\mathbf{u}_t$ to denote $\mathbf{u}_t(M_{1:h})$, the loss function ℓ_t is given by

$$\ell_t(M_{1:h}) = c_t(\mathbf{x}_t, \mathbf{u}_t).$$

Since the cost c_t is a convex function with respect to its arguments, we simply need to show that $\mathbf{x}_t$ and $\mathbf{u}_t$ depend linearly on $M_{1:h}$. The state is given by

$$\mathbf{x}_{t+1} = A\mathbf{x}_t + B\mathbf{u}_t + \mathbf{w}_t = A\mathbf{x}_t + B\left(\sum_{i=1}^{h} M_i \mathbf{w}_{t-i}\right) + \mathbf{w}_t.$$

By induction, we can further simplify:

$$\mathbf{x}_{t+1} = \sum_{i=0}^{t} A^i \left(B \sum_{j=1}^{h} M_j \mathbf{w}_{t-i-j} + \mathbf{w}_{t-i} \right), \tag{7.2}$$

which is a linear function of the variables.

Similarly, the control $\mathbf{u}_t$ is given by

$$\mathbf{u}_t = \sum_{i=1}^{h} M_i \mathbf{w}_{t-i}.$$

Thus, we have shown that $\mathbf{x}_t$ and $\mathbf{u}_t$ are linear transformations of $M_{1:h}$. A composition of convex and linear functions is convex (see Exercise 7.1), which concludes our lemma. □

7.3.2 Loss Functions with Memory

The actual loss c_t at time t is not calculated on $\mathbf{x}_t(M_{1:h})$ but rather on the true state $\mathbf{x}_t$, which in turn depends on different parameters $M_{1:h}^i$ for various historical times $i < t$. However, $c_t(\mathbf{x}_t, \mathbf{u}_t)$ is well approximated by $l_t(M_{1:h}^t)$, as shown next.

Proof of Lemma 7.4 We first note that by the choice of step size by the online gradient descent (Algorithm 6.1), as per Theorem 6.1, we have that $\eta = \frac{D}{G\sqrt{T}}$. Thus,

$$\|M_{1:h}^t - M_{1:h}^{t-i}\| \leq \sum_{s=t-i+1}^{t} \|M_{1:h}^s - M_{1:h}^{s-1}\| \leq i\eta G = \frac{iD}{\sqrt{T}}.$$

We next use this fact to establish that $\mathbf{x}_t$ and $\mathbf{x}_t(M_{1:h}^t)$ are close. Similar to the expansion of $\mathbf{x}_t(M_{1:h})$ in Equation 7.2, $\mathbf{x}_t$ can be written as

$$\mathbf{x}_{t+1} = \sum_{i=0}^{t} A^i \left(B \sum_{j=1}^{h} M_j^{t-i} \mathbf{w}_{t-i-j} + \mathbf{w}_{t-i} \right).$$

Recall that the stability of the time-invariant linear dynamical system $\{A, B\}$ implies that for any i, $\|A^i\| \leq \kappa(1-\delta)^i$, with $\frac{\kappa}{\delta} \leq \gamma$. Therefore, there exists a constant $C \geq 0$ such that

$$\|\mathbf{x}_t - \mathbf{x}_t(M_{1:h}^t)\| = \left\| \sum_{i=0}^{t} A^i B \sum_{j=1}^{h} M_j^{t-i} \mathbf{w}_{t-i-j} - \sum_{i=0}^{t} A^i B \sum_{j=1}^{h} M_j^t \mathbf{w}_{t-i-j} \right\|$$

$$\leq \sum_{i=0}^{t} \|A^i\| \|B\| \sum_{j=1}^{h} \|M_j^{t-i} - M_j^t\| \|\mathbf{w}_{t-i-j}\|$$

$$\leq C \cdot \kappa \sum_{i=0}^{t} (1-\delta)^i \cdot i \frac{hD}{\sqrt{T}}$$

$$= \frac{Ch\kappa D}{\sqrt{T}} \sum_{i=1}^{\infty} i(1-\delta)^{i-1}$$

$$\leq \frac{C\gamma^2 hD}{\sqrt{T}},$$

where on the second last line we use $\sum_{i=1}^{\infty} i(1-\delta)^i = \frac{1}{\delta^2}$.

The control at time t depends only on the parameters at the current iteration and not on historical parameters; hence $\mathbf{u}_t = \mathbf{u}_t(M_{1:h}^t)$. By definition, $l_t(M_{1:h}^t) = c_t(\mathbf{x}_t(M_{1:h}^t), \mathbf{u}_t(M_{1:h}^t))$. Thus, we have

$$l_t(M_{1:h}^t) - c_t(\mathbf{x}_t, \mathbf{u}_t) = c_t(\mathbf{x}_t(M_{1:h}^t), \mathbf{u}_t(M_{1:h}^t)) - c_t(\mathbf{x}_t, \mathbf{u}_t)$$

$$\leq L \cdot \|\mathbf{x}_t(M_{1:h}^t) - \mathbf{x}_t\| \leq \frac{L\gamma^2 ChD}{\sqrt{T}}.$$

□

7.4 Bibliographic Remarks

Online nonstochastic control is an extension of the framework of online convex optimization (Hazan, 2016), as it is applied to policies, and thus to loss functions with memory (Anava et al., 2015).

The bulk of the material in this chapter is from Agarwal et al. (2019c). The work of Agarwal et al. (2019c) defined the general online nonstochastic setting in which the disturbances and the cost functions are adversarially chosen, and where the cost functions are arbitrary convex functions.

The follow-up work by Agarwal et al. (2019b) achieves a logarithmic pseudo-regret for strongly convex, adversarially selected losses, and well-conditioned stochastic noise. Logarithmic regret algorithms for the full setting of online nonstochastic control with strongly convex costs were obtained in Cassel et al. (2020), Foster and Simchowitz (2020), and Simchowitz (2020).

Online nonstochastic control with known dynamics was further extended in several directions. The task of control with bandit feedback for the costs was considered in Gradu et al. (2020b). More refined regret metrics applicable to changing dynamics were studied in Gradu et al. (2020a), Minasyan et al. (2021), Singh (2022), and Zhao et al. (2022). More extensions suitable for changing environments were studied in Baby and Wang (2022). Online nonstochastic control of population dynamics was studied in Golowich et al. (2024).

Li et al. (2021) studies algorithms that balance regret minimization while adhering to constraints. An application of nonstochastic control in iterative planning in the presence of perturbations is described in Agarwal et al. (2021). Various open-source software packages, such as (Gradu et al., 2020a), provide benchmarked implementations of many such algorithms.

An alternate performance metric for online control, which is also nonstochastic, is the competitive ratio. This objective tracks the cost of the online controller and compares it to the best policy in hindsight. Control methods that minimize the competitive ratio were initiated in Goel and Wierman (2019). This approach was extended in a series of papers (Goel et al., 2019; Shi et al., 2020; Goel and Hassibi, 2021). Thus far, the success of this approach was limited to fully observed linear dynamical systems and further to quadratic costs when achieving the optimal competitive ratio. Goel et al. (2023) show that regret minimizers in the non stochastic setting automatically achieve a near-optimal competitive ratio.

A precursor to online nonstochastic control is a line of work, beginning with Abbasi-Yadkori et al. (2014), focused on the control of stochastic systems under sequentially revealed adversarial costs functions. See Cassel and Koren (2020) and Cassel et al. (2022) for further developments.

7.5 Exercises

7.1. Let $l: \mathbb{R}^d \to \mathbb{R}$ be a convex function. Prove that for any matrix $A \in \mathbb{R}^{d \times n}$, $l(Ax)$ is convex for $x \in \mathbb{R}^n$.

7.5 Exercises

7.2.

(i) Let $l: \mathbb{R}^d \to \mathbb{R}$ be a β-smooth function and let $A \in \mathbb{R}^{d \times n}$ be a matrix. Show that $l(A)$ is necessarily a smooth function for $x \in \mathbb{R}^n$.
(ii) Let $l: \mathbb{R}^d \to \mathbb{R}$ be a μ-strongly convex function and let $A \in \mathbb{R}^{d \times n}$ be a matrix. Show that $l(A)$ is not necessarily strongly convex for $x \in \mathbb{R}^n$.

7.3. Prove that the loss functions ℓ_t are convex in $M_{1:h}$, as per Lemma 7.3, even if the initial state is nonzero.

7.4. Consider control of a general, nonlinear dynamical system. Give a version of the GPC (Algorithm 7.1) that can be applied to such a setting (without proving any performance guarantee).

7.5. In this exercise, we prove and extend Theorem 7.2 for linear time-varying systems that are stabilizable, rather than stable.

(i) Write down and prove the theorem for time-varying systems $\{A_t, B_t\}$ that are all γ-stable.
(ii) Consider a sequence of γ-stabilizable time-varying systems $\{A_t, B_t\}$. Show how to reduce this case to the previous one by considering a sequence of linear time-varying γ-stable systems $\tilde{A}_t, B_t$, where $\tilde{A}_t = A_t + B_t K_t$.

8
Online Nonstochastic System Identification

In the previous chapters, we focused on control algorithms that assume knowledge of the system dynamics. However, in many real-world applications, system parameters are unknown or change over time. This necessitates an approach in which the controller must simultaneously learn the system model while making control decisions. This problem, known as system identification, is fundamental to adaptive control, reinforcement learning, and modern data-driven methods for control.

8.1 The Role of System Identification in Online Control

System identification plays a crucial role in control design, as it allows controllers to estimate unknown system parameters in real time. Unlike classical identification methods that rely on extensive offline data collection, an online approach must identify the system while actively controlling it. This presents several unique challenges:

- Feedback loops and nonstationarity: Unlike supervised learning settings, where data is collected passively, control decisions influence future observations, making the estimation process highly dynamic.
- Adversarial or arbitrary disturbances: The disturbances affecting the system may not follow a probabilistic model and can be chosen adversarially, leading to challenges in maintaining robustness.
- Balancing exploration and performance: The controller must explore sufficiently to estimate system parameters accurately while ensuring stable performance, a trade-off that directly impacts regret minimization.

8.2 A Nonstochastic Approach to System Identification

Traditional system identification techniques often rely on stochastic assumptions, where noise is assumed to be drawn from a known distribution, and system parameters are estimated using statistical inference techniques. However, in nonstochastic system identification, no assumptions are made about the noise distribution or system perturbations. Instead, identification strategies are designed to be robust to worst-case disturbances and to operate under adversarial uncertainty.

This shift in perspective has several implications:

- Performance guarantees without probabilistic assumptions: Unlike stochastic estimation methods, which provide guarantees under specific noise models, nonstochastic approaches aim for regret bounds that hold in arbitrary environments.
- Regret as a measure of learning efficiency: The effectiveness of an identification strategy is measured not in terms of statistical efficiency but in its ability to achieve low regret, ensuring that the system is identified quickly enough to enable near-optimal control.
- Computational tractability in identification: Classical system identification methods often involve solving nonconvex optimization problems, which makes them computationally demanding. Nonstochastic approaches frequently leverage convex relaxation techniques to sidestep intractability.

The problem of online system identification is closely related to online convex optimization and sequential decision making. Many regret minimization techniques developed for online learning can be adapted to system identification settings. However, a key distinction is that, in control settings, the learned system model directly affects future decisions, creating a coupling between estimation and control that does not typically exist in standard online learning formulations.

This connection between system identification and online learning leads to several fundamental questions:

- How should a controller optimally trade off exploration (learning system parameters) and exploitation (minimizing control cost)?
- What are the best algorithmic strategies for estimating system parameters while controlling in adversarial settings?
- What regret guarantees can be achieved in online nonstochastic system identification, and how do they compare to traditional learning-based control approaches?

8.3 Nonstochastic System Identification

The problem we study in this section is online nonstochastic system identification of a linear time-invariant dynamical system.

As a first step, we consider the question of whether the system matrices A, B can be recovered from observations over the state, without regard to the actual cost of control.

The method below to achieve this learns the system matrices by using a random control. Specifically, at each step the control $\mathbf{u}_t$ will be randomly selected from a Rademacher distribution, according to Algorithm 8.1.

Algorithm 8.1 System Identification via Method of Moments.

Input: controllability index k, oversampling parameter T_0.
for $t = 0, \ldots, T_0(k+1)$ **do**
    Execute the control $\mathbf{u}_t = \eta_t$ with $\eta_t \sim_{i.i.d.} \{\pm 1\}^d$.
    Record the observed state $\mathbf{x}_t$.
end for
Set $\widehat{A^j B} = \frac{1}{T_0} \sum_{t=0}^{T_0-1} \mathbf{x}_{t(k+1)+j+1} \eta_{t(k+1)}^\top$ for all $j \in \{0, \ldots k\}$.
Define $\widehat{C_0} = (\widehat{B}, \ldots \widehat{A^{k-1}B})$, $\widehat{C_1} = (\widehat{AB}, \ldots \widehat{A^k B})$, and return $\widehat{A}, \widehat{B}$ as

$$\widehat{B} = \widehat{B}, \quad \widehat{A} = \arg\min_{A} \|\widehat{C_1} - A\widehat{C_0}\|_F.$$

The main guarantee from this simple system identification procedure is given as follows.

Theorem 8.1 *Assuming that*

(a) *the system (A, B) is γ-stable,*
(b) *the system is (k, κ)-strongly controllable,*
(c) *the perturbations $\mathbf{w}_t$ do not depend on $\mathbf{u}_t$,*

and for $T_0 = \Omega\left(\frac{d^2 k^2 \kappa^2 \gamma^6}{\varepsilon^2} \log \frac{1}{\Delta}\right)$, the following holds. Algorithm 8.1 outputs a pair $(\widehat{A}, \widehat{B})$ that satisfies, with probability $1 - \Delta$,

$$\|\widehat{A} - A\| \leq \varepsilon, \|\widehat{B} - B\| \leq \varepsilon.$$

The proof of this theorem hinges upon the key observation that

$$\mathbf{E}[\mathbf{x}_{t+k} \mathbf{u}_t^\top] = \mathbf{E}\left[\sum_{i=0:k} A^i (B\mathbf{u}_{t+k-i} + \mathbf{w}_t) \mathbf{u}_t^\top\right] = A^k B$$

8.3 Nonstochastic System Identification

and the crucial fact that the controls are i.i.d. Thus, we can create an unbiased estimator for this matrix, and using many samples, we obtain an increasingly accurate estimate.

Proof First, we establish that $A^j B$ is close to $\widehat{A^j B}$ for all $j \in [k]$.

Lemma 8.2 *As long as $T_0 \geq \tilde{\Omega}\left(\frac{d^2 \gamma^2}{\varepsilon^2} \log \frac{1}{\Delta}\right)$, with probability $1 - \Delta$, we have for all $j \in [k]$ that*

$$\|\widehat{A^j B} - A^j B\| \leq \varepsilon.$$

Proof First note that for any t, since the system is γ stable and the controls have norm $\sqrt{d}$, the state $\mathbf{x}_t$ is bounded as

$$\|\mathbf{x}_t\| = \left\|\sum_{i=1}^{t} A^{i-1}\left(B\eta_{t-i} + \mathbf{w}_{t-i}\right)\right\| \leq 2\sqrt{d}\gamma.$$

Therefore, defining $N_{j,t} = \mathbf{x}_{t(k+1)+j+1}\eta_{t(k+1)}^\top$ for any $j < k$ and $t < T_0$, we have that $\|N_{j,t}\| \leq 2d\gamma$. Note that $\{N_{j,t}\}_{t \leq T_0}$ is not a sequence of independent random variables, since the same instance of η_t might occur in multiple terms. To remedy this, we claim that the sequence $\tilde{N}_{j,t} = N_{j,t} - A^j B$ forms a martingale difference sequence with respect to the filtration Σ_t, defined as $\Sigma_t = \{\eta_i : i < (k+1)(t+1)\}$. Since $N_{j,t}$ only involves the first $t(k+1) + k$ random control inputs, $N_{j,t}$ is Σ_t-measurable. Now, for any t, since η_t is zero-mean and chosen independently of $\{w_j\}, \{\eta_j\}_{j \neq t}$ at each time step, we have

$$\mathbb{E}\left[N_{j,t}|\Sigma_{t-1}\right] = \mathbb{E}\left[\sum_{i>1} A^{i-1}\left(\mathbf{w}_{t(k+1)+j+1-i} + B\eta_{t(k+1)+j+1-i}\right)\eta_{t(k+1)}^\top \bigg| \Sigma_{t-1}\right]$$

$$= \mathbb{E}\left[\sum_{i>1} A^{i-1} B\eta_{t(k+1)+j+1-i}\eta_{t(k+1)}^\top \mathbb{I}_{i=j+1} \bigg| \Sigma_{t-1}\right] = A^j B,$$

where the last equality follows from the fact that $\mathbb{E}[\eta_t \eta_t^\top] = I$.

Theorem 8.3 (Matrix Azuma) *Consider a sequence of random symmetric matrices $\{X_k\} \in \mathbb{R}^{d \times d}$ adapted to a filtration sequence $\{\Sigma_k\}$ such that*

$$\mathbb{E}[X_k|\Sigma_{k-1}] = 0 \quad \text{and} \quad X_k^2 \preceq \sigma^2 I \text{ holds almost surely,}$$

then, for all $\varepsilon > 0$, we have

$$P\left(\lambda_{\max}\left(\frac{1}{K}\sum_{k=1}^{K} X_k\right) \geq \varepsilon\right) \leq d e^{-K\varepsilon^2/8\sigma^2}.$$

Applying the Matrix Azuma inequality (see the bibliography for a reference) on the symmetric dilation, that is,

$$\begin{bmatrix} 0 & \tilde{N}_{j,t}^\top \\ \tilde{N}_{j,t} & 0 \end{bmatrix},$$

we have for all $j \in [k]$ with probability $1 - \Delta$ that

$$\|\widehat{A^j B} - A^j B\| \le 8d\gamma \sqrt{\frac{1}{T_0} \log \frac{n+d}{\Delta}}.$$

A union bound over $j \in [k]$ concludes the claim. □

At this point, define $C_0 = (B, AB, \ldots, A^{k-1}B)$ and $C_1 = (AB, A^2B, \ldots, A^kB)$ to observe that with probability $1 - \Delta$, we have that

$$\|\widehat{C_0} - C_0\|, \|\widehat{C_1} - C_1\| \le \sqrt{k}\varepsilon.$$

By this display, we already have $\|B - \widehat{B}\| \le \varepsilon$. We will make use of this observation to establish the proximity of $\widehat{A}$ and A, using the following lemma, the proof of which is left as Exercise 8.3.

Lemma 8.4 *Let $X, X + \Delta X$ be the solutions to the linear system of equations $XA = B$ and $(X + \Delta X)(A + \Delta A) = B + \Delta B$. Then, as long as $\|\Delta A\| < \sigma_{\min}(A)$, we have*

$$\|\Delta X\| \le \frac{\|\Delta B\| + \|\Delta A\|\|X\|}{\sigma_{\min}(A) - \|\Delta A\|}.$$

Note that $\widehat{A}$ does not necessarily satisfy $\widehat{A}\widehat{C_0} = \widehat{C_1}$. We now use the above lemma on the normal equations, that is, on $AC_0 C_0^\top = C_1 C_0^\top$ and $\widehat{A}\widehat{C_0}\widehat{C_0}^\top = \widehat{C_1}\widehat{C_0}^\top$. First, observe that for any $\varepsilon < 1$,

$$\left\|C_0 C_0^\top - \widehat{C_0}\widehat{C_0}^\top\right\| = \left\|C_0 C_0^\top - C_0 \widehat{C_0}^\top + C_0 \widehat{C_0}^\top - \widehat{C_0}\widehat{C_0}^\top\right\|$$
$$\le \|C_0\|\|C_0 - \widehat{C_0}\| + \|\widehat{C_0}\|\|C_0 - \widehat{C_0}\|$$
$$\le 2\|C_0\|\sqrt{k}\varepsilon + k\varepsilon^2 \le 3\gamma k\varepsilon,$$

where the last line uses that $\|C_0\| \le \gamma$ following the stability of (A, B). A similar display implies the same upper bound on $\|C_1 C_1^\top - \widehat{C_1}\widehat{C_1}^\top\|$. Therefore, using $\|A\| \le \gamma$ and the perturbation result for linear systems, we have

$$\|A - \widehat{A}\| \le \frac{12\gamma^2 k}{\sigma_{\min}(C_0 C_0^\top)}\varepsilon,$$

as long as $6k\gamma\varepsilon < \sigma_{\min}(C_0 C_0^\top)$. Finally, note that (k, κ)-strong controllability implies $\sigma_{\min}(C_0 C_0^\top) \ge 1/\kappa$. □

8.4 From Indentification to Nonstochastic Control

Having recovered approximate estimates $(\widehat{A}, \widehat{B})$ of the system matrices, a natural question arises: how well does the gradient perturbation (Algorithm 7.1) perform on the original system (A, B) when provided with only approximate estimates of the same? We briefly state a result to this extent, whose proof we leave to the explorations of the interested reader.

Lemma 8.5 *Given estimates of the system matrices* $(\widehat{A}, \widehat{B})$ *such that*

$$\|\widehat{A} - A\|, \|\widehat{B} - B\| \leq \varepsilon,$$

the GPC algorithm when run on the original system (A, B) *guarantees a regret bound of*

$$\max_{\mathbf{w}_{1:T}: \|\mathbf{w}_t\| \leq 1} \left(\sum_{t=1}^{T} c_t(\mathbf{x}_t, \mathbf{u}_t) - \min_{\pi \in \Pi^{DAC}} \sum_{t=1}^{T} c_t(\hat{\mathbf{x}}_t, \pi(\hat{\mathbf{x}}_t)) \right) \leq O(\sqrt{T} + \varepsilon T).$$

This result translates to a regret bound of $\tilde{O}(T^{2/3})$ for nonstochastic control of unknown, yet stable, linear systems, by first running the nonstochastic system identification procedure, and subsequently using the estimates thus obtained to run GPC. Details of how to derive this regret bound are left as Exercise 8.1.

8.5 Summary and Key Takeaways

This chapter explored online nonstochastic system identification, where a controller must simultaneously learn an unknown system model while making control decisions in an adversarial setting. Unlike classical stochastic system identification, which assumes structured noise and statistical convergence, the nonstochastic approach makes no such assumptions and instead relies on regret minimization techniques to ensure efficient learning.

The key insights from this chapter are as follows.

- **The Exploration–Exploitation Tradeoff** A fundamental challenge in online system identification is balancing exploration, which enables learning the system dynamics, and exploitation, which minimizes control cost. Too much exploration leads to high immediate costs, while too much exploitation can result in poor long-term learning and suboptimal control performance. The algorithms in this chapter address this by incorporating controlled perturbations in the control inputs, ensuring sufficient exploration while keeping regret low. However, the trade-off results in a regret of $O(T^{2/3})$, rather than the $O(\sqrt{T})$ we had in the last chapter.

More sophisticated techniques for balancing exploration and exploitation can result in better regret bounds, some of which are surveyed in the bibliographic section.
- **Regret Minimization as a Learning Metric** Unlike traditional identification methods that aim at asymptotic convergence, online nonstochastic identification is evaluated in terms of regret, measuring the performance relative to the best fixed model in hindsight.

8.5.1 Looking Ahead

This part of the book considered the online nonstochastic control problem, the regret metric, and online learning techniques applied to control. We have thus far considered the case in which the state is observable to the controller – an assumption that does not always hold.

In the next parts of the book, we consider the setting of partial observability, where the system state is unknown, but an observed signal is available. We start with the problem of learning and move onto online control.

8.6 Bibliographic Remarks

Controlling linear dynamical systems given unknown or uncertain system parameters has long been studied in adaptive control; see the text of Sastry et al. (1990), for example. However, the results from adaptive control typically concern themselves with stability and, even when discussing implications for optimality, present asymptotic bounds.

In contrast, much of the work at the intersection of machine learning and control is focussed on precise nonasymptotic bounds on sample complexity or regret. Fiechter (1997), who studies the discounted LQR problem under **stochastic noise**, is the earliest example that provides a precise sample complexity bound. Similarly, the early work of Abbasi-Yadkori and Szepesvári (2011) uses an optimism-based approach to achieve optimal regret in terms of dependence on horizon. This regret bound was later improved for sparse systems in Ibrahimi et al. (2012). Dean et al. (2018, 2020) established regret and sample complexity results for the general stochastic LQR formulation with a polynomial dependence on natural system parameters. Fazel et al. (2018) studies the efficacy of first-order optimization-based approaches in synthesizing controllers for the LQR problem. The work of Mania et al. (2019) establishes that simple explore-then-commit strategies achieve the optimal dependence of regret on horizon, with Simchowitz and Foster (2020) proving that such approaches are also optimal in terms of their dependence on the dimension of the system. Cohen et al. (2019) also gave a first optimal-in-horizon polynomial regret result simultaneously with the former work, albeit using an optimism-based approach. The result of Plevrakis and Hazan (2020) gives a rate-optimal regret bound for unknown systems with convex costs.

In terms of **nonstochastic** control, the method-of-moments system identification procedure and the corresponding analysis presented here appeared in Hazan et al. (2020), which also gave the first end-to-end regret result for nonstochastic control starting with unknown system matrices. Even earlier, Simchowitz et al. (2019) presented a regression procedure for identifying Markov operators under nonstochastic noise. Both of these approaches have roots in the classical Ho-Kalman system identification procedure (HO and Kálmán, 1966), for which the first nonasymptotic sample complexity, although for stochastic noise, was furnished in Oymak and Ozay (2019). For strongly convex costs, logarithmic regret was established in Simchowitz (2020). Nonstochastic control of unknown, unstable systems without any access to prior information, also called black-box control, was addressed in Chen and Hazan (2021).

The Matrix Azuma inequality used in this chapter appears in Tropp (2012).

8.7 Exercises

8.1. Fix a value of T. Consider running the nonstochastic system identification procedure for $T_0 = \tilde{\Theta}(T^{2/3})$ steps to obtain system matrix estimates $\widehat{A}, \widehat{B}$, and subsequently using these estimates as inputs to the GPC algorithm when executed over the system (A, B). Given Lemma 8.5, conclude that the resultant control procedure produces a regret of at most $\tilde{O}(T^{2/3})$ without any prior knowledge of system matrices, as long as the underlying system is stable.

8.2. Design an analogous control procedure as that for Exercise 8.1 with an identical regret upper bound (up to constants), when the horizon T is not known in advance.

8.3. Prove Lemma 8.4.

PART III

LEARNING AND FILTERING

9
Learning in Unknown Linear Dynamical Systems

In previous chapters, we focused on control in dynamical systems, often assuming that the system state was fully observable and that system dynamics were either known or could be estimated through system identification. However, in many real-world scenarios, these assumptions do not hold. Instead, an agent may only have access to partial observations of the system and must learn to make accurate predictions without directly influencing the system's evolution.

This chapter focuses on learning in a dynamical system, which means predicting future system outputs based on past observations while the underlying system parameters remain unknown. Unlike control, where an agent actively influences system trajectories to minimize a given cost, learning in this setting is a passive task; the goal is to accurately forecast future outputs without modifying the system's behavior.

The ability to predict future states and outputs of an unknown system is fundamental to many domains, including time series forecasting, financial modeling, sensor networks, autonomous systems, and most recently large language models. Moreover, learning a predictive model is often a key stepping stone toward solving more complex problems, such as control under partial observability or reinforcement learning in dynamical systems. In this chapter, we study online learning algorithms for prediction in unknown linear dynamical systems, with an emphasis on regret minimization.

9.1 Learning in Dynamical Systems

The term *learning* in dynamical systems can refer to a variety of tasks, including state estimation, system identification, and policy learning. In this chapter, we focus on the specific task of prediction, where the objective is to estimate future outputs of a dynamical system based on past observations. Unlike classical system identification, which seeks to recover the system parameters explicitly, the learning approach taken here aims to optimize the predictions directly.

To formally define the problem, consider a sequence of observable inputs fed into a dynamical system, as illustrated in Figure 9.1. The learner observes these inputs and aims to predict the corresponding outputs while minimizing a given loss function. For the remainder of this chapter, we consider a partially

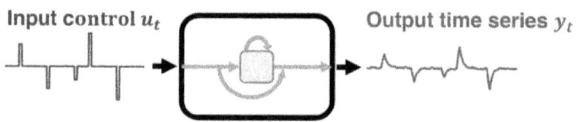

Figure 9.1 In learning we passively observe the input sequence and attempt to predict output.

observable linear time-invariant dynamical system described by the following equations. The extension of the material in this chapter to time-varying linear dynamical systems is left as Exercise 9.6.

$$\mathbf{x}_{t+1} = A\mathbf{x}_t + B\mathbf{u}_t + \mathbf{w}_t,$$
$$\mathbf{y}_t = C\mathbf{x}_t + \mathbf{v}_t.$$

Here, $\mathbf{x}_t$ represents the hidden system state, $\mathbf{u}_t$ is the observed control input, and $\mathbf{y}_t$ is the measured system output. The terms $\mathbf{w}_t$ and $\mathbf{v}_t$ account for the process noise and the observation noise, respectively. A crucial challenge in this setting is that the system matrices A, B, C are unknown to the learner.

The learner's objective is to construct an efficient prediction rule that estimates $\mathbf{y}_t$ based on past observations, despite not knowing the underlying parameters of the system.

9.2 Online Learning of Dynamical Systems

In an online learning setting, the learner sequentially observes the data and makes predictions in real time. Formally, the process unfolds as follows:

(i) At each time step t, the learner observes the control input $\mathbf{u}_{t-1}$.
(ii) The learner makes a prediction $\hat{\mathbf{y}}_t$ for the next output.
(iii) The true output $\mathbf{y}_t$ is revealed.
(iv) The learner incurs a loss based on the difference between $\mathbf{y}_t$ and $\hat{\mathbf{y}}_t$, using a predefined loss function $\ell(\hat{\mathbf{y}}_t, \mathbf{y}_t)$, such as squared error:

$$\ell(\hat{\mathbf{y}}_t, \mathbf{y}_t) = \|\hat{\mathbf{y}}_t - \mathbf{y}_t\|^2.$$

The goal is to minimize the total loss over time. However, in adversarial or unpredictable settings, direct loss minimization may be ill posed. Instead, we consider a regret-based formulation, where the learner's performance is measured relative to the best possible predictor from a predefined class of functions.

More precisely, let Π be a family of prediction rules mapping past observations to predictions. The regret of an online learning algorithm A with respect to Π is given by

$$\text{Regret}_T(A, \Pi) = \sum_{t=1}^{T} \|\mathbf{y}_t - \hat{\mathbf{y}}_t\|^2 - \min_{\pi \in \Pi} \sum_{t=1}^{T} \|\mathbf{y}_t - \hat{\mathbf{y}}_t^\pi\|^2,$$

where $\hat{\mathbf{y}}_t^\pi$ denotes the prediction made at time t by the best function $\pi \in \Pi$ in hindsight.

This regret-based framework allows us to compare different classes of predictors and design learning algorithms that efficiently adapt to unknown dynamical systems.

9.3 Classes of Prediction Rules

We now explore what constitutes a reasonable prediction rule for the output of a linear dynamical system. However, before considering the prediction rules themselves, we discuss how to compare classes of such rules. This is analogous to the definitions and discussion in Chapter 6 where we discussed comparison between policy classes for control. However, the definitions and the corresponding conclusions for making predictions are simpler.

Definition 9.1 We say that a class of predictors Π_1 ε-approximates class Π_2 if the following holds. For every sequence of T inputs and outputs $\{\mathbf{u}_{1:T}, \mathbf{y}_{1:T}\}$, and every $\pi_2 \in \Pi_2$, there exists $\pi_1 \in \Pi_1$ such that

$$\sum_{t=1}^{T} \|\mathbf{y}_t^{\pi_1} - \mathbf{y}_t^{\pi_2}\| \leq T\varepsilon,$$

where $\mathbf{y}_t^{\pi_1} = \pi_1(\mathbf{y}_{1:t-1}, \mathbf{u}_{1:t-1})$, $\mathbf{y}_t^{\pi_2} = \pi_2(\mathbf{y}_{1:t-1}, \mathbf{u}_{1:t-1})$.

The significance of this definition is that if a simple predictor class approximates another more complicated class, then it suffices to consider only the first. The implication of ε-approximation on the regret is given below.

Lemma 9.2 *Let predictor class Π_1 ε-approximate class Π_2. Then for any prediction algorithm $\mathcal{A}$ we have*

$$\text{regret}_T(\mathcal{A}, \Pi_2) \leq \text{regret}_T(\mathcal{A}, \Pi_1) + \varepsilon T.$$

The proof of this lemma is left as an exercise to the reader at the end of this chapter. We now proceed to consider specific prediction rules.

9.3.1 Linear Predictors

The simplest prediction rule is a linear function of observations, inputs, or both. More precisely,

Definition 9.3 (Linear policies) A linear controller parametrized by matrices $\{M_{1,i}, M_{2,j}\} \in \mathbb{R}^{h \times d_y \times d_y + k \times d_y \times d_u}$ is a linear operator mapping state to control as

$$\hat{\mathbf{y}}_t = \sum_{i=1}^{h} M_{1,i} \mathbf{y}_{t-i} + \sum_{j=1}^{k} M_{2,j} \mathbf{u}_{t-j}.$$

We denote by $\Pi^L_{h,k,\kappa}$ the set of all linear controllers bounded by $\sum_{i=1}^{h} \|M_{1,i}\| + \sum_{j=1}^{k} \|M_{2,j}\| \leq \kappa$.

9.3.2 Linear Dynamical Predictors

An intuitive class of prediction rules that makes use of the special structure of linear dynamical systems is that of open-loop linear dynamical predictors, which we now define.

The motivation for this class is the case that the system transformations are known and the system is fully observed, namely $\mathbf{C} = I$, the identity matrix. In this situation, the optimal prediction can be calculated using the system evolution equation $\mathbf{x}_{t+1} = A\mathbf{x}_t + B\mathbf{u}_t$. It can be seen that this prediction rule is optimal for any zero-mean noise or perturbation model. However, this optimality is only true if the state is fully observable. In the next chapter, we consider optimal predictors for zero-mean i.i.d. noise in the more general partially observed case.

9.3 Classes of Prediction Rules

For a partially observed linear dynamical system without noise, we can write

$$\mathbf{y}_t^\pi = \pi_{\{A,B,C,\mathbf{x}_0\}}(t) = C\mathbf{x}_t$$
$$= CA\mathbf{x}_{t-1} + CB\mathbf{u}_{t-1}$$
$$\vdots$$
$$= \sum_{i=1:t-1} CA^{i-1}B\mathbf{u}_{t-i} + CA^t \mathbf{x}_0.$$

Thus, the operator $\pi_{\{A,B,C,\mathbf{x}_0\}}$ takes a history of states and controls and gives the next output of the system accurately. We denote by $\prod^\star$ the set of all such predictors that in the case of zero noise give a perfectly accurate prediction.

If the spectral radius of A is larger than one, then the system is unstable and the magnitude of $\mathbf{y}_t$ can grow arbitrarily large, as discussed in Section 4.2. Thus, for the rest of this section, assume that A is γ-stable.

With a decaying system, we can limit the memory of the predictor and still maintain accuracy. Thus, it makes sense to consider a more restricted class of predictors that have bounded memory, that is,

$$\pi_{\{A,B,C,h\}} = \sum_{i=1}^{h} CA^{i-1} B\mathbf{u}_{t-i},$$

where as a matter of convention we say $\mathbf{u}_t = 0$ for all negative t. Denote by $\prod_h^\star$ the class of all such predictors.

Clearly, for a large h this class of predictors well approximates the true observations $\mathbf{y}_t$ if the system is activated without any noise. We next formalize what this means.

Lemma 9.4 *Suppose that $\|B\| + \|C\| \leq \gamma$, and $\|\mathbf{u}_t\| \leq U$ for all $t \in [T]$. Then the class $\prod_h^\star$ for $h = \gamma \log \frac{T\gamma^3 U}{\varepsilon}$, ε-approximates the class $\prod^\star$.*

Proof Consider any sequence and any predictor from $\prod^\star$, say $\pi_1 = \pi_{A,B,C,\mathbf{x}_0}$. Consider the predictor $\pi_2 = \pi_{A,B,C,h} \in \prod_h^\star$ for the same system matrices. Then we have

$$\|\mathbf{y}_t^{\pi_1} - \mathbf{y}_t^{\pi_2}\|$$

$$= \left\| \sum_{i=1:t-1} CA^{i-1}B\mathbf{u}_{t-i} + CA^{t-1}\mathbf{x}_0 - \sum_{i=1:h} CA^{i-1}B\mathbf{u}_{t-i} \right\|$$

$$= \left\| \sum_{i=h}^{T} CA^{i}B\mathbf{u}_{t-i-1} \right\|$$

$$\leq \sum_{i=h}^{T} \kappa(1-\delta)^{i}\|C\|\|B\|\|\mathbf{u}_{t-i-1}\|$$

$$\leq \frac{\gamma^2 \kappa U}{\delta}(1-\delta)^h \leq \frac{\gamma^2 \kappa U}{\delta} e^{-\delta h},$$

for some $\frac{\kappa}{\delta} \leq \gamma$. Using the value of H, and summing up over all T iterations, we have

$$\sum_t \|\mathbf{y}_t^{\pi_1} - \mathbf{y}_t^{\pi_2}\| \leq \sum_t \frac{\gamma^2 \kappa U}{\delta} e^{-\delta h} = \varepsilon.$$

□

9.4 Efficient Learning of Linear Predictors

The previous section gives several natural classes of predictors as well as some of the relationships between them. In particular, the class of linear dynamical predictors is optimal for certain scenarios and requires knowledge of only three matrices. However, the parameterization of this class is nonconvex in the matrices A, B, C, which presents a computational challenge.

This challenge is circumvented by using the fact that the class $\prod^L$ ε-approximates the class $\prod^\star$. In addition, the natural parameterization of $\prod^L$ is convex and thus can be learned using online convex optimization.

The following algorithm learns the class of linear predictors in the sense that it attains a sublinear regret bound for any sequence of adversarial cost functions, as well as perturbations and observation noise.

This algorithm is an instance of the online gradient descent algorithm, and Theorem 6.1 directly gives the following bound for it. We denote by D the diameter of the set $\mathcal{K}$ as in Algorithm 9.1 and by G an upper bound on the norm of the gradients of the loss functions ℓ_t.

Corollary 9.5 *For choice of step sizes $\eta_t = \frac{D}{G\sqrt{t}}$, and $\mathcal{A}$ being Algorithm 9.1, we have that*

Algorithm 9.1 Learning $\prod^L$ by Online Gradient Descent

1: Input: $M^1 = M^1_{1,1:h}, M^1_{2,1:k}$, convex constraints set $\mathcal{K} \subseteq \mathbb{R}^{h \times d_y + k \times d_u}$
2: **for** $t = 1$ to T **do**
3:    Predict $\hat{\mathbf{y}}_t(M^t) = \sum_{i=1}^{h} M^t_i \mathbf{u}_{t-i} + \sum_{j=1}^{k} M^t_j \mathbf{y}_{t-j}$ and observe true $\mathbf{y}_t$.
4:    Suffer loss $\ell_t(\hat{\mathbf{y}}_t(M^t), \mathbf{y}_t)$, and update:
$$M^{t+1} = M^t - \eta_t \nabla \ell_t(M^t)$$
$$M^{t+1} = \prod_{\mathcal{K}}(M^{t+1})$$
5: **end for**

$$\text{regret}(\mathcal{A}) = \sum_t \ell(\hat{\mathbf{y}}_t^{\mathcal{A}}, \mathbf{y}_t) - \min_{\pi \in \prod^L} \ell(\hat{\mathbf{y}}_t^{\pi}, \mathbf{y}_t) \leq 2GD\sqrt{T}.$$

9.5 Summary

In this chapter, we explored the problem of learning in unknown linear dynamical systems, where the goal is to predict future system outputs based on past observations while the underlying system parameters remain unknown. Unlike traditional control settings where an agent actively influences system behavior, the learning problem here is passive, focusing on accurate forecasting without intervention.

We began by formulating the problem in terms of a partially observable LDS, highlighting the challenges posed by unknown system matrices and noisy observations. We discussed key approaches for online learning in dynamical systems, emphasizing regret minimization as a performance metric. The regret framework allows direct comparisons between learning algorithms by evaluating their cumulative prediction error against the best fixed predictor in hindsight.

A key takeaway from this chapter is that learning in an LDS requires balancing expressivity and computational efficiency. Simple linear predictors may be computationally efficient but lack expressivity, whereas more sophisticated dynamical predictors require careful algorithmic design to remain tractable. The regret analysis provided insights into how online learning algorithms adapt to unknown system parameters while ensuring competitive performance.

This discussion sets the stage for the next chapter, on Kalman filtering, where we transition from open-loop predictors to closed-loop estimation techniques that integrate feedback from observations to refine state estimates.

9.6 Bibliographic Remarks

The task of predicting future observations in an online environment is often called "filtering." Lying at the crossroads between statistical estimation and dynamical systems, the theory of filtering had its early beginnings in the works of Norbert Wiener (see, e.g., Wiener (1949)).

For a comprehensive treatment of online learning in games and online convex optimization, see Cesa-Bianchi and Lugosi (2006) and Hazan (2016).

In the statistical noise setting, a natural approach for learning is to perform system identification and then use the identified system for prediction. This approach was taken by Simchowitz et al. (2018), Simchowitz and Foster (2020), and Sarkar and Rakhlin (2019). The work of Ghai et al. (2020) extends identification-based techniques to adversarial noise and marginally stable systems.

The class $\prod^*$, which is the class of optimal open-loop predictors for systems with full observation, was studied in Hazan et al. (2017). That paper showed how to learn partially observable and even marginally stable dynamical systems whose spectral radius approaches one, as long as their transition matrix is symmetric.

Under probabilistic assumptions, the work of Hardt et al. (2018) shows that the matrices A, B, C, D of an unknown and partially observed time-invariant linear dynamical system can be learned using first-order optimization methods. Learning without recovery was studied in Ghai et al. (2020) even for marginally stable systems. More recently, tensor methods were used in Bakshi et al. (2023a) and in Bakshi et al. (2023b) to learn a mixture of linear dynamical systems.

9.7 Exercises

9.1. Prove that a ε-approximation of prediction classes implies an additional εT regret, as formalized in the Lemma 9.2.

9.2. Prove that the class $\prod^*$ is optimal for fully observed linear dynamical systems with zero mean noise.

9.3. Show that the policy class $\prod_h^\star$ can be learned by the policy class $\prod_{\tilde{h},k,\kappa}^L$ with $\tilde{h} = 0$.

9.4. Give precise regret bounds for learning the class $\prod_{h,k,\kappa}^L$ and $\prod_h^\star$ as a function of the natural class parameters h, k, κ and any other relevant parameter.

9.5. Consider the square loss in Algorithm 9.1. Give an algorithm based on the Online Newton Step algorithm (see Hazan (2016)) and write down the resulting regret bound.

9.6. Write down the predictor class $\prod^\star$ and $\prod_h^\star$ for time-varying linear dynamical systems. Research the notion of adaptive regret from Hazan (2016), and reason about the conditions under which the class $\prod^L$ approximates $\prod^*$ for time-varying linear dynamical systems.

10
Kalman Filtering

State estimation plays a central role in control and prediction tasks, where an agent seeks to infer latent states from noisy observations. The Kalman filter is a classical and widely used method for solving this problem, providing optimal state estimates under the assumption of linear system dynamics and Gaussian noise.

A key feature of the Kalman filter and a fundamental limitation is that it requires knowledge of the system matrices, including both the state transition matrix and the observation model. These matrices encode how the system evolves over time and how observations relate to the hidden state. Although the Kalman filter is provably optimal when these parameters are known, real-world applications often involve partially known or entirely unknown system dynamics, making explicit identification challenging.

In contrast, online learning approaches, such as those introduced in Chapter 9, provide a model-free open-loop alternative by directly learning optimal linear predictors from the data. Instead of estimating a latent state and system matrices, these methods construct forecasting models using past observations, making them more adaptive to unknown or time-varying systems. However, these approaches do not incorporate feedback, meaning that they operate in an open-loop setting, predicting future states without correcting for new observations.

This chapter moves beyond open-loop prediction to closed-loop state estimation, where past observations continuously refine future predictions. The Kalman filter accomplishes this by recursively updating estimates based on incoming measurements, achieving Bayes-optimal performance under Gaussian noise. This structured feedback mechanism allows for more accurate and robust estimation compared to open-loop methods, particularly in dynamic and uncertain environments.

Beyond classical Kalman filtering, we also introduce a closed-loop online learning method that serves as a convex relaxation of the Kalman filter. This approach retains the Kalman filtering feedback mechanism while eliminating the need for explicit system identification, making it applicable in adversarial or uncertain environments. By formulating state estimation as an online convex optimization problem, this method provides provable regret guarantees and serves as a bridge between Bayesian filtering and learning-based estimation techniques.

We provide a rigorous introduction to Kalman filtering, including its derivation, optimality properties, and computational aspects. We also discuss its limitations and contrast it with learning-based alternatives, setting the stage for spectral filtering techniques in the next chapter, which further generalize state estimation in unknown dynamical systems.

10.1 Observable Systems

Observability is a fundamental concept in control theory that determines whether the internal state of a system can be uniquely inferred from its output over time. In the infinite-horizon setting, prediction is meaningful only if observations remain bounded, a condition closely related to system controllability (as discussed in Section 4.3).

A general dynamical system maps an initial state to an infinite sequence of observations. If this mapping is injective, meaning that each initial state corresponds to a unique sequence of observations, then the system is said to be observable. Formally:

Definition 10.1 A partially observed dynamical system is observable if and only if every sequence of observations uniquely determines the initial hidden state.

A simple example of an unobservable system is given as the following variant of the double integrator from Section 2.1.2. Consider the double integrator dynamical system with $\Delta = 1$ and partial observability where the observation matrix is given as follows:

$$A = \begin{bmatrix} 1 & 1 \\ 0 & 1 \end{bmatrix}, \quad C = [0 \ 1].$$

This system allows only the velocity of the object to be observed, while the position remains hidden. It can be seen that this setting is unobservable, as the initial position cannot be determined on the basis of the velocities and forces applied to the system alone. Indeed, this is a special case of the Galilean relativity principle of Newtonian mechanics, which states that the laws of physics are the same in all inertial reference frames.

For linear systems, observability can be verified using a rank condition on the Kalman observability matrix, given by

$$\hat{K}(A, C) = \begin{bmatrix} C \\ CA \\ \cdots \\ CA^{d_x-1} \end{bmatrix},$$

where A and C are the system matrices that define the state transition and observation model, respectively. Notice that in our example, the Kalman observability matrix is given by $\hat{K}(A, C) = \begin{bmatrix} 0 & 1 \\ 0 & 1 \end{bmatrix}$, which is degenerate. This is not by chance; the system is observable if and only if the observability matrix has full rank. We ask the reader to provide a proof of the following result in Exercise 10.1.

Theorem 10.2 *A partially-observed dynamical system specified by system parameters (A, B, C) is observable if and only if* $\mathrm{rank}(\hat{K}(A, C)) = d_x$.

10.2 Optimality of the Kalman Filter

10.2.1 The Power of Linear Predictors

One of the important conclusions from this chapter is that the policy class of linear predictors is optimal with respect to closed-loop policies. We prove this by proving two facts:

(i) The class of linear policies ε-approximates the class of Kalman filters, in the sense of Definition 9.1.
(ii) The class of Kalman filters is optimal for closed-loop policies.

The conclusion from these two facts is that algorithm 9.1 can be used to learn a policy that competes with the best closed-loop optimal predictor in hindsight. If the signal comes from a linear dynamical system subject to Gaussian noise, this means competing with the optimal predictor, which is a Kalman filter. We prove this formally at the end of this section.

10.2.2 Closed-Loop Optimal Predictors

Open-loop predictors, such as those introduced in the previous chapter, leverage system inputs and initial conditions to make predictions. However, they do not

10.2 Optimality of the Kalman Filter

incorporate feedback from observations, making them susceptible to modeling errors and disturbances.

Closed-loop prediction rules address this limitation by continuously adjusting estimates based on new observations. The most well-known example of such a predictor is the Kalman filter, which produces the best linear predictor in a least-squares sense under the assumption that process and observation noise are Gaussian with known covariances.

Formally, let $\mathbf{w}_t$ and $\mathbf{v}_t$ denote the process and observation noise, assumed to be zero-mean and i.i.d., with covariance matrices

$$\Sigma_x \stackrel{\text{def}}{=} \mathbb{E}\left[\mathbf{w}_t \mathbf{w}_t^\top\right], \quad \Sigma_y = \mathbb{E}\left[\mathbf{v}_t \mathbf{v}_t^\top\right].$$

So far, we have focused on predicting observations, that is, $\mathbf{y}_t$. However, the Kalman filter also provides optimal state estimates $\hat{\mathbf{x}}_t$ given past observations $\mathbf{y}_{t-1:1}$ and controls $\mathbf{u}_{t-1:1}$. This is motivated by the fact that the least-squares optimal predictor of observations can be expressed as

$$\mathbf{y}_t = C\mathbf{x}_t.$$

In Exercise 10.6, we ask the reader to provide a proof of this statement.

Based on this observation, the following theorem is the essence of the Kalman filter, which recovers estimates for states as well as linear projection matrices for optimal prediction of the next observation.

Theorem 10.3 *Let the control sequence $\mathbf{u}_t$ be an arbitrary affine function of $\mathbf{y}_{1:t}$. There exists an efficiently computable sequence of linear predictors $\hat{\mathbf{x}}_{t+1}$ defined recursively as*

$$\hat{\mathbf{x}}_{t+1} = (A - L_t C)\hat{\mathbf{x}}_t + B\mathbf{u}_t + L_t \mathbf{y}_t,$$

such that

$$\mathbb{E}\|\mathbf{x}_t - \hat{\mathbf{x}}_t\|^2 \leq \min_{L \in \prod^L} \mathbb{E}\|\mathbf{x}_t - L[\mathbf{y}_{1:t-1}, \mathbf{u}_{1:t-1}]\|^2.$$

To prove this theorem, we require some basic tools from linear regression, which we state next.

10.2.3 Tools from Linear Regression

We begin with an observation about least-squares predictors, which we ask the reader to prove in Exercise 10.3.

Lemma 10.4 *Let X, Y, Z be matrix-valued random variables. Let*

$$A^* = \arg\min_A \mathbb{E}\|X - AY\|^2, \text{ and } B^* = \arg\min_B \mathbb{E}\|X - BZ\|^2.$$

If $\mathbb{E}[YZ^\top] = 0$, then

$$\mathbb{E}\|X - A^*Y - B^*Z\|^2 \le \min_{A,B} \mathbb{E}\|X - AY - BZ\|^2.$$

In the remaining proof, we use this observation to construct an optimal least-squares estimator $\hat{\mathbf{x}}_{t+1}$ for $\mathbf{x}_{t+1}$ given $\hat{\mathbf{x}}_t$. In particular, we know by induction hypothesis that

$$\mathbb{E}\|\mathbf{x}_t - \hat{\mathbf{x}}_t\|^2 \le \min_{L \in \Pi^L} \mathbb{E}\|\mathbf{x}_t - L[\mathbf{y}_{1:t-1}, \mathbf{u}_{1:t-1}]\|^2.$$

But, is $\hat{\mathbf{x}}_t$ any good for predicting $\mathbf{x}_{t+1}$? To answer this, we note the following structural characterization of least-squares (Exercise 10.2).

Lemma 10.5 *Let X, Y be possibly correlated random variables. Then*

$$A^* \stackrel{\text{def}}{=} \arg\min_A \mathbb{E}\|X - AY\|^2 = \mathbb{E}[XY^\top]\left(\mathbb{E}[YY^\top]\right)^{-1}.$$

Furthermore, if $\hat{X} = A^*Y$, then $\mathbb{E}(X - \hat{X})Y^\top = 0$.

10.2.4 Analysis

We are now ready to prove the main property of the Kalman filter below.

Proof of Theorem 10.3 Since the estimator is defined recursively, a natural strategy is to proceed by induction. Let us say $\hat{\mathbf{x}}_t$ is such a predictor. Furthermore, let

$$\Sigma_t \stackrel{\text{def}}{=} \mathbb{E}(\mathbf{x}_t - \hat{\mathbf{x}}_t)(\mathbf{x}_t - \hat{\mathbf{x}}_t)^\top.$$

Since for any $s < t$,

$$\mathbb{E}\left[(\mathbf{x}_{t+1} - B\mathbf{u}_t)\mathbf{y}_s^\top\right] = \mathbb{E}\left[(A\mathbf{x}_t + \mathbf{w}_t)\mathbf{y}_s^\top\right] = A\mathbb{E}\left[\mathbf{x}_t \mathbf{y}_s^\top\right],$$

using the first part of Lemma 10.5, we have that

$$\mathbb{E}\|\mathbf{x}_{t+1} - A\hat{\mathbf{x}}_t - B\mathbf{u}_t\|^2 \le \min_{L \in \Pi^L} \mathbb{E}\|\mathbf{x}_{t+1} - B\mathbf{u}_t - L[\mathbf{y}_{1:t-1}, \mathbf{u}_{1:t-1}]\|^2.$$

But, when predicting $\mathbf{x}_{t+1}$, we have more information. In particular, we also have $\mathbf{y}_t$. Consider random variables $\mathbf{y}_t - C\hat{\mathbf{x}}_t$. Can we use this covariate alone to predict $\mathbf{x}_{t+1}$? Note that

$$\mathbb{E}\left[(\mathbf{y}_t - C\hat{\mathbf{x}}_t)(\mathbf{y}_t - C\hat{\mathbf{x}}_t)^\top\right] = \Sigma_v + C\mathbb{E}\left[(\mathbf{x}_t - \hat{\mathbf{x}}_t)(\mathbf{x}_t - \hat{\mathbf{x}}_t)^\top\right]C^\top$$
$$= \Sigma_v + C\Sigma_t C^\top,$$

10.2 Optimality of the Kalman Filter

$$\mathbb{E}\left[(\mathbf{x}_{t+1} - B\mathbf{u}_t)(\mathbf{y}_t - C\hat{\mathbf{x}}_t)^\top\right] = \mathbb{E}\left[(A\mathbf{x}_t + \mathbf{w}_t)(C(\mathbf{x}_t - \hat{\mathbf{x}}_t) + \mathbf{v}_t)^\top\right]$$
$$= A\mathbb{E}\left[(\mathbf{x}_t - \hat{\mathbf{x}}_t)(\mathbf{x}_t - \hat{\mathbf{x}}_t)^\top\right]C^\top + A\underbrace{\mathbb{E}\left[\hat{\mathbf{x}}_t(\mathbf{x}_t - \hat{\mathbf{x}}_t)^\top\right]}_{=0}C^\top$$
$$= A\Sigma_t C^\top,$$

where the equality indicated in braces follows from the second part of Lemma 10.5, by noting that $\hat{\mathbf{x}}_t$ is a linear function of $\mathbf{y}_{1:t-1}$.

In the preceding paragraphs, we have constructed two least-squares predictors for $\hat{\mathbf{x}}_{t+1}$. Observe that for any $s < t$,

$$\mathbb{E}(\mathbf{y}_t - C\hat{\mathbf{x}}_t)\mathbf{y}_s^\top = C\mathbb{E}(\mathbf{x}_t - \hat{\mathbf{x}}_t)\mathbf{y}_s^\top = 0,$$

where the last equality once again follows from Lemma 10.5. This orthogonality allows us to invoke Lemma 10.4 to arrive at

$$\hat{\mathbf{x}}_{t+1} = A\hat{\mathbf{x}}_t + B\mathbf{u}_t + \underbrace{A\Sigma_t C^\top \left(\Sigma_y + C\Sigma_t C^\top\right)^{-1}}_{\overset{\text{def}}{=} L_t}(\mathbf{y}_t - C\hat{\mathbf{x}}_t),$$

where the last component follows from Lemma 10.5 and the correlations we just computed.

To establish that such predictors can be computed efficiently, note that

$$\mathbf{x}_{t+1} - \hat{\mathbf{x}}_{t+1} = (A - L_t C)(\mathbf{x}_t - \hat{\mathbf{x}}_t) + \mathbf{w}_t + L_t \mathbf{v}_t.$$

Finally, we give a recursive relation on Σ_t as follows:

$$\Sigma_{t+1} \overset{\text{def}}{=} \mathbb{E}\left[(\mathbf{x}_{t+1} - \hat{\mathbf{x}}_{t+1})(\mathbf{x}_{t+1} - \hat{\mathbf{x}}_{t+1})^\top\right]$$
$$= (A - L_t C)\Sigma_t(A - L_t C)^\top + \Sigma_x + L_t \Sigma_y L_t^\top$$
$$= A\Sigma_t A^\top + \Sigma_x + L_t(C\Sigma_t C^\top + \Sigma_y)L_t^\top - 2A\Sigma_t C^\top L_t^\top$$
$$= A\Sigma_t A^\top - A\Sigma C^\top(C\Sigma_t C^\top + \Sigma_y)^{-1}C\Sigma A^\top + \Sigma_x.$$

□

10.2.5 Infinite Horizon Kalman Filtering

In the infinite-horizon setting, the steady-state Kalman filter can be described by the following linear dynamical system:

$$\hat{\mathbf{x}}_{t+1} = (A - LC)\hat{\mathbf{x}}_t + B\mathbf{u}_t + L\mathbf{y}_t,$$
$$\hat{\mathbf{y}}_t = C\hat{\mathbf{x}}_t,$$

where $L = A\Sigma(C\Sigma C^\top + \Sigma_y)^{-1}$, and Σ satisfies the Riccati equation

$$\Sigma = A\Sigma A^\top - A\Sigma C^\top(C\Sigma C^\top + \Sigma_y)^{-1}C\Sigma A^\top + \Sigma_x.$$

But when does a solution to this Riccati equation exist? A sufficient condition for this is the observability of the underlying dynamical system. In particular, any observable partially-observed linear dynamical system admits a solution to the Riccati equation listed above, and hence has a steady-state Kalman filter. Mirroring the link between controllability and stabilizability, observability implies $\rho(A - LC) < 1$.

10.2.6 Linear Predictors and the Kalman Filter

For observable linear dynamical systems, the Kalman filter is contained in the class of linear predictors. This is formally stated in the following lemma, whose proof is left as Exercise 10.4.

Lemma 10.6 *Consider any partially-observed linear dynamical system (A, B, C) where the Kalman filtering matrix L is such that the system $A - LC$ is γ-stable. The Kalman filtering policy is ε-contained in the class of linear policies $\Pi_{h,k,\kappa}^L$, where $h = k = \gamma \log \frac{\gamma T}{\varepsilon}$ and $\kappa = \gamma^3$.*

10.3 Bayes-Optimality under Gaussian Noise

In this subsection, we further restrict $\mathbf{w}_t, \mathbf{v}_t$ to be i.i.d. draws from Gaussian distributions and proceed to establish that the Kalman filter in such cases enjoys a stronger notion of optimality; it produces the best estimator (under a least-squares loss) of the state given past observations among the class of all (not just linear) predictors.[1] The setting we consider is that of Gaussian perturbations $\mathbf{w}_t$ and $\mathbf{v}_t$ and known systems matrices (A, B, C). In particular, we consider the case where $\mathbf{w}_t \sim \mathcal{N}(0, \Sigma_x)$ and $\mathbf{v}_t \sim \mathcal{N}(0, \Sigma_y)$ are i.i.d. Gaussian and $x_0 = 0$. Given the realizations $\mathbf{y}_1, \mathbf{y}_2 \ldots, \mathbf{y}_t$ under some control input sequence $\mathbf{u}_1, \mathbf{u}_2, \ldots, \mathbf{u}_t$, we would like to compute a prediction $\hat{\mathbf{y}}_{t+1}$ that minimizes the mean square error, which is simply the expected observation as per

$$\hat{\mathbf{y}}_{t+1} = \mathbb{E}[\mathbf{y}_{t+1} | \mathbf{y}_{1:t}, \mathbf{u}_{1:t}].$$

Since $\hat{\mathbf{y}}_{t+1} = C\hat{\mathbf{x}}_{t+1}$, we will recursively compute $\hat{\mathbf{x}}_{t+1}$ where

$$\hat{\mathbf{x}}_{t+1} = \mathbb{E}[\mathbf{x}_{t+1} | \mathbf{y}_{1:t}, \mathbf{u}_{1:t}].$$

Theorem 10.7 *The conditional expectation of the state $\hat{\mathbf{x}}_{t+1}$ satisfies*

$$\hat{\mathbf{x}}_{t+1} = (A - L_t C)\hat{\mathbf{x}}_t + B\mathbf{u}_t + L_t \mathbf{y}_t,$$

---

[1] This is also called the Bayes-optimal predictor. See the bibliographic materials for more details and literature.

for a sequence of matrices $L_t \in \mathbb{R}^{d_x \times d_y}$ that can be efficiently computed given (A, B, C), and is independent of the realizations $\mathbf{u}_{1:T}, \mathbf{y}_{1:T}$.

Proof From the Gaussian nature of perturbations, we observe that

$$\begin{bmatrix} \mathbf{x}_t | (\mathbf{y}_{1:t-1}, \mathbf{u}_{1:t-1}) \\ \mathbf{y}_t | (\mathbf{y}_{1:t-1}, \mathbf{u}_{1:t-1}) \end{bmatrix} \sim \mathcal{N}\left(\begin{bmatrix} \hat{\mathbf{x}}_t \\ C\hat{\mathbf{x}}_t \end{bmatrix}, \begin{bmatrix} \Sigma_t & \Sigma_t C^\top \\ C\Sigma_t & C\Sigma_t C^\top + \Sigma_y \end{bmatrix} \right)$$

for some $\Sigma_t \in \mathbb{R}^{d_x \times d_x}$, for which we provide a recursive formula toward the end of the proof. We make note of the following lemma for Gaussian distributions, whose proof we ask the reader to complete in Exercise 10.5.

Lemma 10.8 *If a random variable pair* $(\mathbf{x}, \mathbf{y})$ *are distributed as*

$$\begin{bmatrix} \mathbf{x} \\ \mathbf{y} \end{bmatrix} \sim \mathcal{N}\left(\begin{bmatrix} \mu_x \\ \mu_y \end{bmatrix}, \begin{bmatrix} \Sigma_{11} & \Sigma_{12} \\ \Sigma_{21} & \Sigma_{22} \end{bmatrix} \right),$$

then $\mathbf{x}|\mathbf{y} \sim \mathcal{N}(\mu_x + \Sigma_{12}\Sigma_{22}^{-1}(\mathbf{y} - \mu_y), \Sigma_{11} - \Sigma_{12}\Sigma_{22}^{-1}\Sigma_{21})$.

Therefore, we have that

$$\mathbf{x}_t | (\mathbf{y}_{1:t}, \mathbf{u}_{1:t-1}) \sim \mathcal{N}(\hat{\mathbf{x}}_t + \Sigma_t (C\Sigma_t C^\top + \Sigma_y)^{-1} (\mathbf{y}_t - C\hat{\mathbf{x}}_t), \Sigma_t'),$$
$$\Sigma_t' = \Sigma_t - \Sigma_t C^\top (C\Sigma_t C^\top + \Sigma_y)^{-1} C\Sigma_t.$$

Now, using this, we arrive at

$$\hat{\mathbf{x}}_{t+1} = A\mathbb{E}[\mathbf{x}_t | \mathbf{y}_{1:t}, \mathbf{u}_{1:t-1}] + B\mathbf{u}_t$$
$$= \underbrace{(A - A\Sigma_t (C\Sigma_t C^\top + \Sigma_y)^{-1} C)}_{\stackrel{\text{def}}{=} L_t} \hat{\mathbf{x}}_t + B\mathbf{u}_t + \underbrace{A\Sigma_t (C\Sigma_t C^\top + \Sigma_y)^{-1}}_{\stackrel{\text{def}}{=} L_t} \mathbf{y}_t.$$

It remains to obtain a recursive relation on Σ_t. To this extent, from the state evolution of a linear dynamical system, we have that

$$\Sigma_{t+1} = A\Sigma_t' A^\top + \Sigma_x = A\Sigma_t A^\top - A\Sigma_t C^\top (C\Sigma_t C^\top + \Sigma_y)^{-1} C\Sigma_t A^\top + \Sigma_x.$$

$\square$

10.4 Conclusion

In this chapter, we explored the Kalman filter as a fundamental tool for state estimation in dynamical systems, particularly in the presence of noise and uncertainty. We established its optimality in the least-squares sense and examined its connections to online convex optimization, reinforcing its relevance in both classical control and modern learning-based paradigms.

110   Kalman Filtering

Key insights from this chapter include the role of observability in ensuring accurate state estimation, the derivation of the Kalman filter update equations, and the recursive structure that allows for efficient real-time implementation. We also presented the filter's steady-state behavior, particularly in the infinite-horizon setting, where it converges to a solution governed by the Riccati equation.

Beyond its theoretical guarantees, the Kalman filter is widely used in practical applications ranging from autonomous systems and robotics to finance and neuroscience. However, challenges remain, particularly in scenarios with nongaussian noise, time-varying dynamics, or adversarial disturbances. These limitations motivate extensions such as robust filtering, nonlinear filtering (e.g., the extended and unscented Kalman filters), and learning-based adaptations that integrate statistical inference with modern machine learning techniques.

The next chapter will transition from the classical filtering framework to spectral filtering, providing an alternative approach to state estimation that leverages spectral methods for efficient learning in dynamical systems.

10.5 Bibliographic Remarks

In 1960, Rudolph Kalman gave a state-space solution to the filtering problem that was highly amenable to further extensions (Kalman, 1960). See Anderson and Moore (1991) for a historical perspective on its development and aftermaths.

For unknown systems, in the stochastic case, one may estimate the system matrices, as we have discussed before, before applying optimal filtering. Improper learning approaches to this problem were studied in Kozdoba et al. (2019).

State estimation is a fundamental problem in control and signal processing, with numerous approaches beyond the classical Kalman filter. Alternative methods include:

- **Nonlinear Filtering Extensions** The Kalman filter assumes that the dynamics is a linear system and the noise is Gaussian, which limits its effectiveness in nonlinear settings. The extended Kalman filter (see, e.g., Urrea and Agramonte (2021)) addresses this issue by linearizing the dynamics at each time step via a first-order Taylor-series expansion. However, linearization can introduce significant errors in highly nonlinear systems. Notable extensions using more sophisticated sampling include the unscented Kalman filter

(Julier and Uhlmann, 2004), leading to better estimation accuracy in strongly nonlinear systems.
- **Particle Filtering (Sequential Monte Carlo Methods)**
  For highly nonlinear and non-Gaussian settings, particle filters (Gordon et al., 1993) provide a flexible alternative. These methods approximate the posterior distribution using a set of weighted samples (particles) and employ importance sampling and resampling techniques.
- **Adversarial & Robust Filtering**
  Robust filtering methods seek to provide guarantees under adversarial noise. The H_∞ filter (Başar and Bernhard, 2008) minimizes the worst-case estimation error rather than the mean-squared error, making it suitable for applications with unknown disturbances.

Although the Kalman filter remains a cornerstone of state estimation, these alternative methods provide valuable extensions and improvements for nonlinear, adversarial, and data-driven settings. In the next chapter, we will study an alternative using a technique that circumvents the need for state estimation altogether and has stronger guarantees for sequential prediction.

10.6 Exercises

10.1. In this exercise we prove Theorem 10.2 in two parts:

(i) Show that if the rank of the Kalman observability matrix is less than $d_\mathbf{x}$, then there are two initial states that after $d_\mathbf{x}$ iterations map to the same state. Hence, the initial state is not uniquely determined.
(ii) In contrast, show that if the rank of $\hat{K}$ is full, then at any time t, $\mathbf{x}_t$ determines $\mathbf{x}_0$ uniquely.

10.2. Prove Lemma 10.5, which states the following. Let X, Y be possibly correlated random variables. Then

$$A^* \stackrel{\text{def}}{=} \arg\min_A \mathbb{E}\|X - AY\|^2 = \mathbb{E}\left[XY^\top\right] \left(\mathbb{E}\left[YY^\top\right]\right)^{-1}.$$

Furthermore, if $\hat{X} = A^*Y$, then $\mathbb{E}(X - \hat{X})Y^\top = 0$.

10.3. Using Lemma 10.5, prove Lemma 10.4, which states the following. Let X, Y, Z be matrix-valued random variables. Let

$$A^* = \arg\min_A \mathbb{E}\|X - AY\|^2, \text{ and } B^* = \arg\min_B \mathbb{E}\|X - BZ\|^2.$$

If $\mathbb{E}\left[YZ^\top\right] = 0$, then

$$\mathbb{E}\|X - A^*Y - B^*Z\|^2 \leq \min_{A,B} \mathbb{E}\|X - AY - BZ\|^2.$$

10.4. In this exercise we conclude that the Kalman filter can be improperly learned by the class of linear predictors. Prove this fact as formally stated in Lemma 10.6. Conclude that the Kalman filter can be learned using Algorithm 9.1.

10.5. Prove Lemma 10.8, which is stated below. If a random variable pair $(\mathbf{x}, \mathbf{y})$ are distributed as

$$\begin{bmatrix} \mathbf{x} \\ \mathbf{y} \end{bmatrix} \sim \mathcal{N}\left(\begin{bmatrix} \mu_x \\ \mu_y \end{bmatrix}, \begin{bmatrix} \Sigma_{11} & \Sigma_{12} \\ \Sigma_{21} & \Sigma_{22} \end{bmatrix} \right),$$

then the conditional distribution of $\mathbf{x}$ given $\mathbf{y}$ can be written as $\mathbf{x}|\mathbf{y} \sim \mathcal{N}(\mu_x + \Sigma_{12}\Sigma_{22}^{-1}(\mathbf{y} - \mu_y), \Sigma_{11} - \Sigma_{12}\Sigma_{22}^{-1}\Sigma_{21})$.

10.6. Using Lemma 10.5, prove that the optimal least-squares estimator for the observation at time t given past observations $y_{1:t-1}$ and control inputs $u_{1:t-1}$ is given by $\hat{\mathbf{y}}_t = C\hat{\mathbf{x}}_t$, where $\hat{\mathbf{x}}_t$ is as defined in Theorem 10.3.

11
Spectral Filtering

In the previous chapter, we explored two major approaches for state estimation and prediction in unknown linear dynamical systems:

(i) State estimation via Kalman filtering, which recursively estimates both the latent system state and the projection matrices that map the system state to observations. Kalman filtering provides an optimal solution under Gaussian noise and updates the state estimates over time. However, it requires an accurate model of system dynamics, including transition and observation matrices, which may be difficult to estimate, especially in adversarial or high-noise environments.

(ii) Online learning of linear predictors, which offers a relaxation of the Kalman filtering approach by directly learning predictive models without explicit system identification. Instead of estimating the full state and transition matrices, this approach constructs a forecasting model using past inputs and outputs. Although this avoids the complexity of full system identification, it introduces a major limitation: The number of parameters required depends on the stability constant of the system, which can be large and unknown, making learning impractical for unstable or poorly conditioned systems.

In this chapter, we introduce spectral filtering, an alternative approach that bypasses these limitations. Rather than relying on state estimation or direct system identification, spectral filtering constructs predictors on a transformed basis, which does not depend on the stability properties of the system. This method leverages spectral decompositions to efficiently extract the relevant dynamical structure, enabling more robust and scalable learning in both stable and unstable systems.

To illustrate the power of spectral filtering, we first present a one-dimensional case, which provides an intuitive understanding of how spectral methods can be applied to system identification and forecasting. We then extend the discussion to higher-dimensional settings, demonstrating how spectral predictors enable a structured and computationally efficient representation of system dynamics.

Although this chapter primarily focuses on symmetric system matrices for clarity, the principles of spectral filtering extend to asymmetric and more general linear dynamical system settings. These extensions require additional mathematical tools, which we reference in the bibliographic section.

11.1 Spectral Filtering in One Dimension

In this section, we give a simple introduction to the technique of spectral filtering, restricting ourselves to a single dimension. The restriction to a single dimension simplifies the exposition and yet is sufficient to illustrate the main ideas. In the next section, we consider the high-dimensional case.

The one-dimensional systems that we consider next are scalar systems, with $d_u = d_y = d_x = 1$. For simplicity, we assume that the initial state is zero, $x_0 = 0$. Recall the set of optimal predictors, the class $\Pi_\star$, which for scalar systems can written and simplified to

$$\mathbf{y}_t = \sum_{i=1}^{t} CA^{i-1} B\mathbf{u}_{t-i} = \beta \sum_{i=1}^{t} \alpha^{i-1} \mathbf{u}_{t-i},$$

where $\beta = CB \in \mathbb{R}$. Thus, we can write the difference between two consecutive iterates as

$$\begin{aligned}
\mathbf{y}_t - \mathbf{y}_{t-1} &= \beta \left(\sum_{i=1}^{t} \alpha^{i-1} \mathbf{u}_{t-i} - \sum_{i=1}^{t-1} \alpha^{i-1} \mathbf{u}_{t-1-i} \right) \\
&= \beta \left(\mathbf{u}_{t-1} + \sum_{i=1:t-1} (\alpha^i - \alpha^{i-1}) \mathbf{u}_{t-1-i} \right) \\
&= \beta \mathbf{u}_{t-1} + \beta \cdot \mu_\alpha(\mathbf{u}_{t-1:1}),
\end{aligned} \qquad (11.1)$$

which is a geometric series involving α multiplying the past controls. We used the operator notation to denote the operator

$$\mu_\alpha(\mathbf{u}_{t-1:1}) = \sum_{i=1:t-1} (\alpha - 1)\alpha^{i-1} \mathbf{u}_{t-1-i} = \mu_\alpha^T \tilde{\mathbf{u}}_{t-1}. \qquad (11.2)$$

Above we used the vector notation for

$$\mu_\alpha = (\alpha - 1)\begin{bmatrix} 1 & \alpha & \alpha^2 & \ldots & \alpha^{T-1} \end{bmatrix} \in \mathbb{R}^T,$$

and we define $\tilde{\mathbf{u}}_t$ to be the padded vector $\mathbf{u}_{t-1:1}$:

$$\tilde{\mathbf{u}}_t = \begin{bmatrix} \mathbf{u}_{t-1} & \ldots & \mathbf{u}_1 & 0 & \ldots & 0 \end{bmatrix} \in \mathbb{R}^T.$$

11.1 Spectral Filtering in One Dimension

The main idea of spectral filtering is to rewrite the vectors μ_α in a more *efficient* basis, as opposed to the standard basis. The notion of efficiency comes from the definition of ε-approximation that we have studied in previous chapters. To define the basis, we consider the following matrix:

$$Z_T = \int_0^1 \mu_\alpha \mu_\alpha^\top d\alpha \in \mathbb{R}^{T \times T}.$$

The matrix Z_T is symmetric and is independent of α. Let $\phi_1, ..., \phi_k, ... \in \mathbb{R}^T$ be the eigenvectors of Z_T ordered by the magnitude of their corresponding eigenvalues. Our basis for representing μ_α is spanned by the eigenvectors of Z_T, and we proceed to study their properties.

11.1.1 The Magic of Hankel Matrices

The matrix Z_T is a special case of a **Hankel Matrix**, which is a square Hermitian matrix with each ascending skew-diagonal from left to right having the same value. An example of a Hankel matrix is the following:

$$W_T = \int_0^1 \begin{bmatrix} 1 & \alpha & \alpha^2 & \\ \alpha & \alpha^2 & \alpha^3 & \\ \alpha^2 & \alpha^3 & \alpha^4 & \\ & & & \ddots \\ & & & & \alpha^{2T-2} \end{bmatrix} d\alpha = \begin{bmatrix} 1 & \frac{1}{2} & \frac{1}{3} & \\ \frac{1}{2} & \frac{1}{3} & \frac{1}{4} & \\ \frac{1}{3} & \frac{1}{4} & \frac{1}{5} & \\ & & & \ddots \\ & & & & \frac{1}{2T-1} \end{bmatrix}.$$

This matrix is closely related to the Z_T matrix we have introduced before, but not equal to it.

The properties of Hankel matrices have been studied extensively, and the following theorem captures one of their crucial properties.

Theorem 11.1 *Let σ_k be the kth largest singular value of a Hankel matrix $M \in \mathbb{R}^{T \times T}$, then there exist constants c_1, c_2 such that*

$$\sigma_k(M) \leq c_1 \cdot e^{-\frac{c_2 k}{\log T}}.$$

The proof of this theorem is beyond our scope, and references to details are given in the bibliographic material at the end of this chapter.

This theorem shows that the eigenvalues of a Hankel matrix decay very rapidly. An additional illustration of this fact is given in Figure 11.1, which plots the eigenvalues of the matrix Z_T for $T = 30$ on a logarithmic scale.

Spectral Filtering

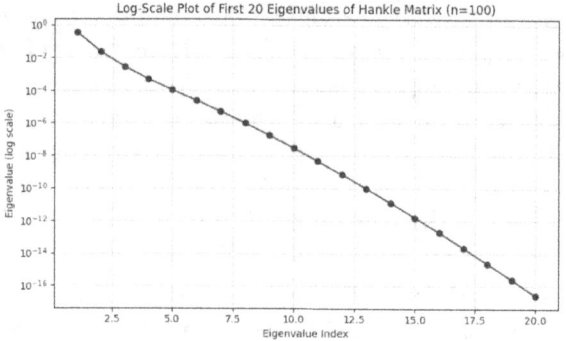

Figure 11.1 Eigenvalues of Hankel matrices decrease geometrically. These are the eigenvalues of the matrix Z_T, where $T = 30$, plotted on a logarithmic scale. Some of the eigenvectors are plotted in Figure 11.2.

11.2 Spectral Predictors

The basis of eigenvectors of a Hankel matrix gives rise to an interesting class of predictors. This class of mappings does not directly apply to the history of inputs and/or observations. Instead, we first apply a linear transformation over these vectors and then apply a linear map. More formally,

Definition 11.2 Define the class of spectral linear predictors of order k recursively as

$$\prod_h^\lambda = \left\{ \pi_M \;\middle|\; \pi_M(\mathbf{u}_{1:t}, \mathbf{y}_{1:t}) - \pi_M(\mathbf{u}_{1:t-1}, \mathbf{y}_{1:t-1}) = M_0 \mathbf{u}_t + \sum_{i=1}^h M_i \cdot \phi_i(\tilde{\mathbf{u}}_{t-1}) \right\}.$$

Here the operator $\phi_i(\mathbf{u}_{t-1})$ takes the inner product of the ith eigenvector of Z_T with $\tilde{\mathbf{u}}_{t-1}$, where $\tilde{\mathbf{u}}_t = \begin{bmatrix} \mathbf{u}_{t-1} & \ldots & \mathbf{u}_1 & 0 & \ldots & 0 \end{bmatrix} \in \mathbb{R}^T$.

Notice that as $h \mapsto T$, this class is equivalent to the class of all linear predictors on the inputs $\prod^{LIN}$. The reader may wonder at this point: Since a linear transformation composed with another linear transform remains linear, what is the benefit here?

The answer is that linear transformations are equivalent when we have full-rank transforms. In our case, however, we consider low-rank approximations to the transform, and the approximation properties of different subspaces can differ drastically! This is the case in our setting: The spectral basis arising from the matrix Z_T gives excellent approximations to the class of (standard) linear predictors, as we formally prove next. The following lemma is analogous to Lemma 9.4, with the significant difference that the number of parameters h does not depend on the stabilizability parameter of the system γ at all!

11.2 Spectral Predictors

Lemma 11.3 *Suppose that $x_0 = 0$, then the class $\prod_h^\lambda$ for $h = O(\log^2 \frac{T}{\varepsilon})$ ε-approximates the class $\prod_\star$.*

The proof of this lemma relies on the following crucial property of the spectral basis, whose proof we defer until later.

Lemma 11.4 *Let $\{\phi_j\}$ be the eigenvectors of Z. Then there exist constants $c_1, c_2 > 0$ such that for all $j \in [T]$, $\alpha \in [0, 1]$, we have*

$$|\phi_j^\top \mu_\alpha| \leq c_1 e^{-\frac{c_2 j}{\log T}}.$$

Using this property, we can prove Lemma 11.3 by simply shifting to a representation on the spectral basis as follows.

Proof of Lemma 11.3 Consider any sequence and any predictor from $\prod_\star$, say $\pi_1 = \pi_{A,B,C,0}$. Following Equation (11.1), we can write the optimal predictor in the spectral basis:

$$\begin{aligned} \mathbf{y}_t^{\pi_1} - \mathbf{y}_{t-1}^{\pi_1} - \beta \mathbf{u}_t &= \beta \cdot \mu_\alpha(\mathbf{u}_{t-1:1}) \\ &= \beta \cdot \mu_\alpha^\top \tilde{\mathbf{u}}_{t-1} \quad &(11.2) \\ &= \beta \cdot \mu_\alpha^\top \left(\sum_{i=1}^T \phi_i \phi_i^\top \right) \tilde{\mathbf{u}}_{t-1} \quad \text{basis property } \sum_{i=1}^T \phi_i \phi_i^\top = I \\ &= \sum_{i=1}^T M_i^\star \cdot \phi_i^\top \tilde{\mathbf{u}}_{t-i}. \quad & M_i^\star = \beta \phi_i^\top \mu_\alpha \end{aligned}$$

By Lemma 11.4 we have that

$$M_i^\star = \beta \phi_i^\top \mu_\alpha = O(e^{-\frac{i}{\log T}}).$$

We proceed by an inductive claim on $\|\mathbf{y}_t^{\pi_1} - \mathbf{y}_t^{\pi_2}\|$. Consider the predictor $\pi_2 \in \prod_h^\lambda$ for the same system matrices $M^\star$. Then we have

$$\|\mathbf{y}_t^{\pi_1} - \mathbf{y}_t^{\pi_2}\| - \|\mathbf{y}_{t-1}^{\pi_1} - \mathbf{y}_{t-1}^{\pi_2}\|$$

$$\leq \|\mathbf{y}_t^{\pi_1} - \mathbf{y}_{t-1}^{\pi_1} - \mathbf{y}_t^{\pi_2} + \mathbf{y}_{t-1}^{\pi_2}\| \qquad \Delta\text{-inequality}$$

$$= \left\|\sum_{i=1}^T M_i^\star \phi_i^\top \tilde{\mathbf{u}}_{t-i} - \sum_{i=1}^h M_i^\star \phi_i^\top \tilde{\mathbf{u}}_{t-i}\right\|$$

$$= \left\|\sum_{i=h+1}^T M_i^\star \phi_i^\top \tilde{\mathbf{u}}_{t-1}\right\|$$

$$\leq \sum_{i=h+1}^T \|M_i^\star\| |\phi_i^\top \tilde{\mathbf{u}}_{t-1}|$$

$$\leq \sum_{i=h+1}^T \|M_i^\star\| \|\phi_i\|_2 \|\tilde{\mathbf{u}}_{t-1}\|_2 \qquad \text{Cauchy–Schwartz}$$

$$\leq \sqrt{T} \cdot \sum_{i=h+1}^T |M_i^\star| \qquad |\mathbf{u}_t| \leq 1, \|\phi_i\|_2 \leq 1$$

$$\leq \sqrt{T} \times O\left(\int_h^\infty e^{-\frac{i}{\log T}} di\right) \qquad \text{Lemma 11.4}$$

$$\leq \frac{\varepsilon}{T}. \qquad \text{choice of } h$$

Thus, since the same argument exactly applies to $\|\mathbf{y}_{t-1}^{\pi_1} - \mathbf{y}_{t-1}^{\pi_2}\|$, we have that

$$\|\mathbf{y}_t^{\pi_1} - \mathbf{y}_t^{\pi_2}\| \leq \sum_{t=1}^T \frac{\varepsilon}{T} \leq \varepsilon.$$

□

It remains to prove Lemma 11.4, which is the main technical lemma of this section.

Lemma 11.4 Consider the scalar function $g(\alpha) = (\mu_\alpha^\top \phi_j)^2$ over the interval $[0, 1]$. First, notice that by the definition of ϕ_j as the eigenvectors of Z_T and σ_j as the corresponding eigenvalues, we have

$$\int_{\alpha=0}^1 g(\alpha) d\alpha = \int_{\alpha=0}^1 (\phi_j^\top \mu_\alpha)^2 d\alpha = \int_{\alpha=0}^1 \phi_j^\top \mu_\alpha \mu_\alpha^\top \phi_j d\alpha = \phi_j^\top Z_T \phi_j = \sigma_j.$$

In addition, we claim that the function $g(\alpha)$ is L-Lipschitz for a constant value of L. To see this, first notice that

$$\|\mu_\alpha\|_2^2 = \sum_{i=0}^{T-1} (1-\alpha)^2 \alpha^{2i} \leq \frac{(1-\alpha)^2}{1-\alpha^2} = \frac{1-\alpha}{1+\alpha} \leq 1-\alpha,$$

11.2 Spectral Predictors

implying that $g(\alpha) = (\mu_\alpha^\top \phi_j)^2 \leq \|\phi_j\|^2 \|\mu_\alpha\|^2 \leq 1 - \alpha$. Next, we have

$$\left\|\frac{\partial}{\partial \alpha}\mu_\alpha\right\|_2^2 = \sum_{t=0}^{T-1}((1-\alpha)i\alpha^{i-1} - \alpha^i)^2$$

$$\leq 2(1-\alpha)^2 \sum_{t=0}^{T-1} i^2 \alpha^{2i-2} + 2\sum_{t=1}^{T} \alpha^{2i} \qquad (a+b)^2 \leq 2(a^2+b^2)$$

$$\leq 2(1-\alpha)^2 \left(\frac{1}{(1-\alpha^2)^2} + \frac{\alpha^2}{(1-\alpha^2)^3}\right) + \frac{2}{1-\alpha^2}$$

$$\leq \frac{6}{1-\alpha},$$

where we use the identity $\sum_{i=0}^\infty i^2 \beta^{i-1} = \frac{1}{(1-\beta)^2} + \frac{\beta}{(1-\beta)^3}$ with the substitution $\beta = \alpha^2$. Using this, we obtain an upper bound on the Lipschitz constant of $g(\alpha)$ as

$$|g'(\alpha)| = \left|\frac{\partial}{\partial \alpha}(\mu_\alpha^\top \phi_j)^2\right| = 2|\mu_\alpha^\top \phi_j|\left|\frac{\partial}{\partial \alpha}(\mu_\alpha^\top \phi_j)\right|$$

$$\leq 2\|\mu_\alpha\|_2\|\phi_j\|_2 \left|\frac{\partial}{\partial \alpha}(\mu_\alpha^\top \phi_j)\right| \qquad \text{Cauchy–Schwarz}$$

$$\leq 2\|\mu_\alpha\|_2\|\phi_j\|_2^2 \left\|\frac{\partial}{\partial \alpha}\mu_\alpha\right\|_2 \qquad \text{Cauchy–Schwarz}$$

$$\leq 2\sqrt{1-\alpha}\sqrt{\frac{6}{1-\alpha}} \leq 6. \qquad \|\phi_i\|_2 \leq 1$$

Let $g_{\max} = \max_\alpha g(\alpha)$ be the maximal value that g attains from zero to one. Any nonnegative L-Lipschitz function that reaches $g_{\max}$ and has area A, satisfies

$$A \geq \frac{g_{\max}^2}{2L} \Rightarrow g_{\max}^2 \leq 2LA,$$

since the minimal area is realized by a triangle.

Thus, the maximum value of $g(\alpha)$ is bounded, using Theorem 11.1 by

$$g_{\max} = \max_\alpha \left\{(\phi_j^\top \mu_\alpha)^2\right\} \leq 2 \cdot 6 \cdot \int_\alpha (\phi_j^\top \mu_\alpha)^2 d\alpha$$

$$= 12 \cdot \phi_j^\top \left(\int_\alpha \mu_\alpha \mu_\alpha^\top d\alpha\right) \phi_j$$

$$= 12\sigma_j$$

$$\leq 12c_1 e^{-\frac{c_2 j}{\log T}}. \qquad \text{Theorem 11.1}$$

□

11.2.1 Generalization to High Dimensions

Consider a linear dynamical system with matrices A, B, C, and further assume that A is symmetric. This implies that A can be diagonalized over the set of reals as $A = U^\top DU$. Decomposing over each dimension $j \in [d]$ separately, and denoting $D_{jj} = \alpha_j$, we can write (11.1):

$$\mathbf{y}_t - \mathbf{y}_{t-1} - CB\mathbf{u}_{t-1}$$

$$= \sum_{i=1}^{t-1} CU^\top (D^{i-1} - D^{i-2}) UB \mathbf{u}_{t-1-i}$$

$$= \sum_{i=1}^{t-1} CU^\top \left(\sum_{j=1}^{d_x} (\alpha_j^{i-1} - \alpha_j^{i-2}) e_j e_j^\top \right) UB \mathbf{u}_{t-i-1}$$

$$= \sum_{j=1}^{d_x} c_j b_j^\top \sum_{i=1}^{t-1} (\alpha_j^{i-1} - \alpha_j^{i-2}) \mathbf{u}_{t-1-i} \qquad c_j, b_j \leftarrow j\text{th row/col of } CU^\top, UB$$

$$= \sum_{j=0}^{d_x} \left(M_j^\star \mathbf{u}_t + M_j^\star \mu_{\alpha_j}(\mathbf{u}_{1:t-1}) \right)$$

$$= \tilde{M}^\star \mathbf{u}_t + \tilde{M}^\star \mu_{\alpha_j}(\mathbf{u}_{1:t-1}).$$

Thus, we can apply the same approximation techniques from the previous section and derive an analogous conclusion to Lemma 11.3 for dimensions higher than one. Crucially, the number of parameters required does not depend on the hidden dimension of the original system. The details are left as Exercise 11.2.

11.3 Online Spectral Filtering

The prediction class $\prod_h^\lambda$ can be learned by online gradient descent (Algorithm 6.1) in the online convex optimization framework, which gives rise to the following algorithm.

The eigenvectors of Z have an interesting structure: Their magnitude does not decay with time, which explains their improved approximation as captured formally in Lemma 11.3.

The regret guarantees of the online gradient descent algorithm, as given in Theorem 6.1, directly imply the following corollary.

Theorem 11.5 *Let D be the diameter of the parameter set $\mathcal{K}$, and let G be an upper bound on the norm of the gradients for the loss functions f_t as in Algorithm 11.1. For choice of step sizes $\eta_t = \frac{D}{G\sqrt{t}}$, we have that*

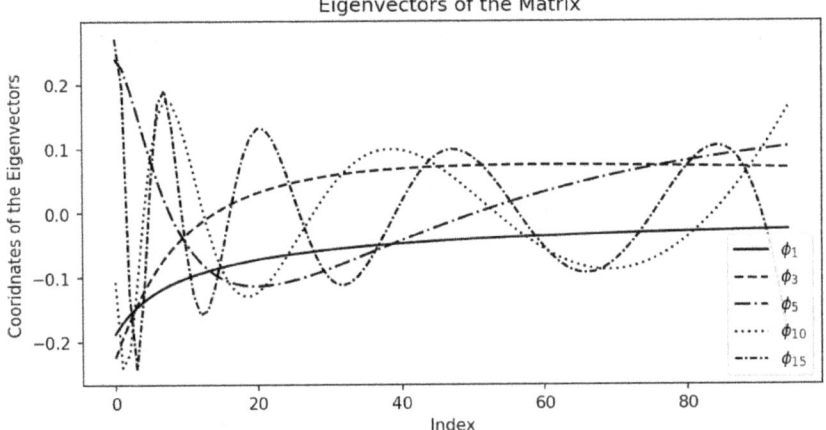

Figure 11.2 The filters obtained by the eigenvectors of Z.

Algorithm 11.1 Online Spectral Filtering

1: Input: $M_{1:h}^1$, convex constraints set $\mathcal{K} \subseteq \mathbb{R}^{h \times d_y \times d_u}$, horizon T
2: Compute h top eigenvectors of Z_T, denoted $\phi_1, ..., \phi_h$.
3: **for** $t = 1$ to T **do**
4:     Predict $\hat{\mathbf{y}}_t = M_0^t \mathbf{u}_t + \sum_{j=1}^h M_j^t \phi_j(\tilde{\mathbf{u}}_{t-i})$, as per Definition 11.2, and observe true $\mathbf{y}_t$.
5:     Define cost function $f_t(M) = \|\mathbf{y}_t - \tilde{\mathbf{y}}_t\|^2$.
6:     Update and project:

$$M^{t+1} = M^t - \eta_t \nabla f_t(M^t)$$
$$M^{t+1} = \prod_{\mathcal{K}}(M^{t+1})$$

7: **end for**

$$\sum_t \|\mathbf{y}_t - \hat{\mathbf{y}}_t\|^2 - \min_{\pi \in \Pi_h^\lambda} \|\mathbf{y}_t - \mathbf{y}_t^\pi\|^2 \leq 2GD\sqrt{T}.$$

11.4 Conclusion

In this chapter, we introduced spectral filtering as a novel approach to state estimation and prediction in linear dynamical systems, offering an alternative to classical methods like Kalman filtering and online learning of linear predictors. Using spectral decompositions, we bypass the need for explicit system identification and provide a framework that is robust to the stability properties of the system.

We began by outlining the core challenges of traditional state estimation approaches. Kalman filtering, while optimal under Gaussian noise assumptions, requires knowledge of system matrices, which may be difficult to estimate in high-noise or adversarial settings. Online learning of linear predictors relaxed these requirements but introduced a dependence on system stability constants, which can be large and unknown. Spectral filtering circumvents these limitations by constructing predictors in a transformed basis, allowing for efficient representation without requiring explicit system matrices.

We demonstrated the method first in the one-dimensional case, providing an intuition for how spectral predictors extract relevant dynamical structure. This was then extended to higher-dimensional settings, where spectral filtering enables a structured and computationally efficient approach to forecasting. Crucially, we showed that this approach does not rely on system stability constants, making it particularly appealing for learning in both stable and unstable dynamical systems.

From a methodological standpoint, we connected spectral filtering to convex optimization techniques, showing that it can be implemented via online learning in a spectral basis. This formulation allows for efficient sequential learning and prediction with provable performance guarantees.

Looking ahead, spectral filtering opens the door for several exciting directions. Extensions to asymmetric system matrices, as referenced in the bibliographic section, provide a pathway toward broader applicability. Moreover, integrating spectral methods with control strategies may lead to novel approaches for nonstochastic control with partial observations, which we explore in the Chapter 12

11.5 Bibliographic Remarks

Deviating from classical recovery and into the world of improper prediction, the work of Hazan et al. (2017) devised an improper learning technique using the eigenvectors of the Hankel matrix Z. The methods we surveyed in previous chapters for learning linear dynamical systems had statistical and computational complexity that depended on the hidden dimension of the system. In contrast, the spectral filtering technique studied in this chapter has regret bounds and computational complexity guarantees that do not depend on the hidden dimension.

The spectral properties of the Hankel matrices used in this chapter are from Beckermann and Townsend (2017).

The first extension of the spectral learning technique to asymmetric matrices is due to Hazan et al. (2018). A more recent advancement of spectral filtering for general linear dynamical systems was recently put forth in Marsden and Hazan (2025).

The technique of spectral filtering has found applications in control and sequence prediction since its introduction.

In Arora et al. (2018), the authors use a spectral filtering technique to concisely represent linear dynamical systems in a linear basis, formulating the optimal control problem as a convex program. This method circumvents the nonconvex challenges associated with explicit system identification and latent state estimation, providing provable optimality guarantees for the control signal.

A recent line of work makes use of the spectral basis discovered in the line of work surveyed in this chapter to design neural architectures for sequence prediction. This was first suggested in the context of state space models in the spectral basis in Agarwal et al. (2023) and was extended to hybrid models in Liu et al. (2024). Later extensions explored efficient implementation (Agarwal et al., 2024) and the concept of length generalization in sequence prediction (Marsden et al., 2024).

11.6 Exercises

11.1. Generalize Lemma 11.3 to all real numbers in the range $[-1, 1]$, instead of $[0, 1]$.

11.2. Generalize Lemma 11.3 to any dimension $d_x > 1$.

11.3. The fact that Z_T is positive semidefinite is important to ensure that the eigenvalues of Z_T decay. Give an example of a family of $n \times n$ nonpositive semidefinite Hankel matrices that have eigenvalues with constant absolute value.

11.4. The view of Theorem 13.5 is a bit misleading, as it compares the iterates of online gradient descent with the hypothesis class Π_h^λ and uses big-O notation to obscure the diameter and Lipschitz constants of the predictors. Assume that $d_x = 1$.

(i) Find a convex set $\mathcal{M}$ of tuples $(M_1, M_2, \dots)$ to optimize over such that its diameter is some polylog$(T, 1/\varepsilon)$. The norm we measure diameter with respect to is the sum of the Frobenius norms of the M_i's.
(ii) In addition, show that the norm of the gradient is at most polylog$(T, 1/\varepsilon)$ as well. Using these two facts, show that with the right value of ε, the regret against the hypothesis class $\Pi_\star$ is at most $\tilde{O}(\sqrt{T})$.

PART IV

ONLINE CONTROL WITH PARTIAL OBSERVATION

12
Policy Classes for Partially Observed Systems

In previous chapters, we explored control and learning techniques in fully observed linear dynamical systems, where the system state is directly accessible to the controller. However, in many real-world scenarios, such as robotics, economics, and signal processing, the state is only partially observed, requiring the controller to infer system dynamics from noisy or incomplete observations.

This chapter focuses on partially observed linear dynamical systems (LDSs), where the state is hidden, and only a linear projection of it is observed. For the remainder of this part of the text, we restrict ourselves to time-invariant systems and refer the reader to the bibliographic section for extensions. The system follows the general form

$$\mathbf{x}_{t+1} = A\mathbf{x}_t + B\mathbf{u}_t + \mathbf{w}_t,$$
$$\mathbf{y}_t = C\mathbf{x}_t.$$

Here, $\mathbf{x}_t$ represents the hidden system state, $\mathbf{u}_t$ the control input, $\mathbf{y}_t$ the observed output, and $\mathbf{w}_t$ the disturbance. Often, partially observed LDSs are defined with an additional noise term that corrupts the observations. Exercise 12.2 asks the reader to prove that even in such circumstances we may assume the previously stated form for partially observed systems without loss of any generality.

Unlike in the fully observed setting, where controllers can react based on the full state, partial observability introduces fundamental limitations on policy design.

In the partial observation setting, some of the control policy classes from Chapter 6 are not well defined. We then define policy classes and describe the relationship between them.

Not all control policies designed for fully observed systems extend naturally to partially observed settings. In this chapter, we systematically explore different policy classes, analyzing their expressivity, computational tractability, and suitability for partially observed LDSs.

- Linear observation–action controllers (or simply linear controllers): These controllers directly map observations to actions but are highly inexpressive, as they ignore past observations and fail to capture hidden state dependencies.
- Generalized linear observation–action controllers (or generalized linear controllers): These extend linear controllers by incorporating the history of observations, significantly improving expressiveness.
- Linear dynamic controllers (LDCs): These policies model internal system dynamics, making them more powerful than simple observation-based policies.
- Disturbance response controllers (DRCs): Introduced as an alternative to state-based policies, DRCs act directly on a counterfactual signal, "Nature's y's," which captures the response of the system that results from uniformly setting the control input to zero. We show that DRCs can approximate LDCs, indicating that they are a highly expressive and yet convex policy class. A schematic relationship between the classes is depicted in Figure 12.1.

We conclude by analyzing the relationships between these policy classes, showing how they fit into the broader framework of nonstochastic control with partial observation. This chapter lays the groundwork for online learning and regret minimization in partially observed LDSs, which we will explore in the next chapter.

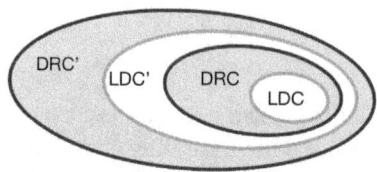

Figure 12.1 Schematic relationship between linear disturbance-response control (DRC) policies and linear dynamical control (LDC) policies of increasing dimensions (or number of parameters) for linear time-invariant systems.

12.1 Linear Observation Action Controllers

Linear observation–action controllers are the most natural generalization of linear controllers to the partially observed case, and are defined as follows.

12.1 Linear Observation Action Controllers

Definition 12.1 A linear observation–action controller $K \in \mathbb{R}^{d_u \times d_y}$ is a linear operator mapping observation to control by $\mathbf{u}_t^\pi = K\mathbf{y}_t$. We denote by Π_γ^L the set of all linear controllers bounded in the Fronbenius norm by $\|K\| \leq \gamma$ that are also γ-stabilizing for given LDS.

Although this class of policies is natural to consider, it turns out to be very inexpressive. Even for optimal control with stochastic noise, the optimal policy cannot be approximated by a linear observation–action policy (see Exercise 12.6). Therefore, we now define more expressive and useful policies. However, linear controllers will be useful in simplifying proofs later on.

12.1.1 Generalized Linear Observation–Action Controllers

We can generalize linear observation–action controllers to apply to a history of past observations. This additional generality, of taking a history of observations, greatly increases the expressive power of this policy class. The formal definition is given below.

Definition 12.2 Generalized linear observation–action controllers parameterized by matrices $K_{0:h}$ choose a control input by

$$\mathbf{u}_t^\pi = \sum_{i=0}^{h} K_i \mathbf{y}_{t-i}.$$

By $\Pi_{h,\gamma}^G$ we denote the class of generalized linear observation–action controllers that satisfy $\sum_{i=0}^{h} \|K_i\| \leq \gamma$ and are also γ-stabilizing.

Generalized observation–action policies can be seen as a special case of observation–action controllers using the technique of dimension lifting as follows. Consider any partially observed system (A, B, C) and perturbation sequence $\mathbf{w}_t$. Consider the *lifted* system $(\tilde{A}, \tilde{B}, \tilde{C}, \tilde{D})$ given by

$$\mathbf{z}_{t+1} = \tilde{A}\mathbf{z}_t + \tilde{B}\mathbf{u}_t + \tilde{D}\mathbf{w}_t, \quad \tilde{\mathbf{y}}_t = C\mathbf{z}_t,$$

where

$$\tilde{A} = \begin{array}{|c|c|c|c|c|} \hline A & 0 & \cdots & 0 & 0 \\ \hline I & 0 & \cdots & 0 & 0 \\ \hline 0 & I & \cdots & 0 & 0 \\ \hline \vdots & \vdots & \cdots & \vdots & \vdots \\ \hline 0 & 0 & \cdots & I & 0 \\ \hline \end{array}, \quad \tilde{B} = \begin{array}{|c|} \hline B_t \\ \hline 0 \\ \hline 0 \\ \hline \vdots \\ \hline 0 \\ \hline \end{array},$$

$$\tilde{C} = \begin{array}{|c|c|c|c|} \hline C & C & \cdots & C \\ \hline \end{array}, \quad \tilde{D} = \begin{array}{|c|} \hline I \\ \hline 0 \\ \hline 0 \\ \hline \vdots \\ \hline 0 \\ \hline \end{array}.$$

The following lemma relates the original and lifted system (for a proof, see Exercise 12.3).

Lemma 12.3 *For any sequence of control inputs* $\mathbf{u}_t$, *it holds that* $\tilde{\mathbf{y}}_t = [\mathbf{y}_t, \mathbf{y}_{t-1}, \ldots \mathbf{y}_{t-h}]$.

The above lemma shows that generalized linear observation–action controllers are a special case of simple observation–action controllers such that $\mathbf{u}_t^\pi = [K_0 \ldots K_h] \tilde{\mathbf{y}}_t$ for the lifted system.

12.2 Linear Dynamic Controllers

The most well-known class of controllers for partially observable linear dynamical systems is that of linear dynamical controllers, due to their connection to Kalman filtering.

Definition 12.4 (Linear dynamic controllers for partially observed systems) A linear dynamic controller π has parameters (A_π, B_π, C_π) and chooses a control at time t as

$$\mathbf{u}_t^\pi = C_\pi \mathbf{s}_t, \quad \mathbf{s}_t = A_\pi \mathbf{s}_{t-1} + B_\pi \mathbf{y}_t.$$

We denote by Π_γ^{LDC} the class of all LDC that are γ-stabilizing as per Definition 2.4 and Exercise 4.5, with parameters $\frac{\kappa}{\delta} \leq \gamma$ that satisfy

$$\forall i, \ \|A_\pi^i\| \leq \kappa(1-\delta)^i, \ (\|B_\pi\| + \|C_\pi\|) \leq \kappa.$$

Note that LDCs are ε-approximated by generalized observation–action controllers. The proof is left as Exercise 12.4.

Lemma 12.5 *Consider a partially observed time-invariant linear dynamical system* $\{A, B, C\}$. *The class* $\Pi^G_{h',\gamma'}$ ε-*approximates the class* Π^{LDC}_γ *for* $h' = O\left(\gamma \log \frac{\gamma}{\varepsilon}\right), \gamma' = 2\gamma^3$.

12.3 Disturbance Response Controllers

The most important policy class in the context of nonstochastic control is that of disturbance response controllers, or DRCs. These policies act on a signal called **Nature's y's**, denoted $\mathbf{y}^{\text{nat}}_t$, which is the would-be observation at time t assuming that all inputs to the system were zero from the beginning of time.

For ease of exposition, we only consider time-invariant linear dynamical systems that are intrinsically stable. This allows us to ignore the stabilizing controller component of disturbance response control and greatly simplifies the presentation of the main ideas. The extension to time-varying systems is explored in the exercises.

Definition 12.6 (Disturbance response controller)   A disturbance response policy for stable systems $\pi_{M_{1:h}}$ is parameterized by matrices $M_{0:h} = [M_0, \ldots, M_h]$. It outputs control $\mathbf{u}^\pi_t$ at time t according to the rule

$$\mathbf{u}^\pi_t = \sum_{i=0}^{h} M_i \mathbf{y}^{\text{nat}}_{t-i},$$

where $\mathbf{y}^{\text{nat}}_t$ is given by the system response to the zero control at time t, that is,

$$\mathbf{x}^{\text{nat}}_{t+1} = A\mathbf{x}^{\text{nat}}_t + \mathbf{w}_t = \sum_{i=0}^{t} A^i \mathbf{w}_{t-i}, \quad \mathbf{y}^{\text{nat}}_t = C\mathbf{x}^{\text{nat}}_t.$$

Denote by $\Pi^R_{h,\gamma}$ the set of all disturbance response policies that satisfy[1]

$$\sum_{i=0}^{h} \|M_i\| \leq \gamma.$$

If the perturbations are all zero, seemingly DRC policies produce zero control. Therefore, we could add a constant offset to the control that can be incorporated as in Definition 6.5.

The main advantage of members of this policy class, as opposed to previously considered policies in optimal control, is that they yield convex parametrization.

---

[1] Recall that for this section, we restrict our discussion to stable systems. Otherwise, we would also ask the policy class to be γ-stabilizing, which can, for example, be achieved by superimposing a fixed γ-stabilizing linear dynamical controller.

Thus, this policy class allows for provably efficient optimization and algorithms, which we explore in the next chapter.

We conclude with the following important lemma on an expression that is important to compute nature's y's. The proof of this lemma follows directly from the definition and is left as Exercise 12.1.

Lemma 12.7 *The sequence $\{\mathbf{y}_t^{\text{nat}}\}$ can be iteratively computed using only the sequence of observations*

$$\mathbf{y}_t^{\text{nat}} = \mathbf{y}_t - C \sum_{i=1}^{t} A^i B \mathbf{u}_{t-i}.$$

Alternatively, the above computation can be carried out recursively as stated below, starting with $\mathbf{z}_0 = \mathbf{0}$:

$$\mathbf{z}_{t+1} = A\mathbf{z}_t + B\mathbf{u}_t,$$
$$\mathbf{y}_t^{\text{nat}} = \mathbf{y}_t - C\mathbf{z}_t.$$

We proceed to show that the class of DRCs approximates the class of LDCs.

12.3.1 Expressivity of Disturbance Response Controllers

A DRC always produces controls that ensure that the system remains state-bounded. This can be shown analogously to Lemma 6.6 (see Exercise 12.5).

Although DRCs are contained within the policy class of DACs, they are still very expressive. Note that a DAC policy cannot be executed in a partially observed system from observations alone; even the knowledge of the entire observation sequence $\mathbf{y}_t$ may not uniquely determine $\mathbf{w}_t$.

We have already seen that LDCs are approximately captured by generalized linear controllers in Lemma 12.5. It can be shown that the class of DRCs with appropriate parameters ε-approximates the class of generalized linear policies for linear time-invariant systems. This observation in conjunction with Lemma 12.5 implies that DRCs also approximate LDCs.

To demonstrate that DRCs approximate generalized linear controllers, we instead prove that DRCs approximate linear controllers, a weaker class of policies. However, this is sufficient to conclude the stronger claim. To see this, first recall that generalized linear controllers acting on the last h states can be written as linear controllers on a lifted system with h states, as noted in Lemma 12.3. Together with the next lemma, this implies that a DRC on the lifted system can approximate generalized linear controllers on the original system. However, this is not quite what we promised. To wrap up the argument, one can verify

12.3 Disturbance Response Controllers

that a DRC with history h' on the lifted system can be written as a DRC with history $h' + h$ on the original system.

Lemma 12.8 *Consider a γ-stable time-invariant linear dynamical system (A, B, C). The class $\Pi_{h',\gamma'}^R$ ε-approximates the class Π_γ^L for the parameters $h' = O\left(\gamma \log \frac{\gamma}{\varepsilon}\right)$, $\gamma' = O(\gamma^6)$.*

Recall that for fully observed LDS, the state–action pairs produced by execution of a linear policy can be written as linear functions of the perturbations. This observation allowed us to prove that DACs capture linear policies for fully observed systems. We prove an analogue for partially observed systems: The observation–action pairs produced by the execution of linear policies can be written as linear functions of nature's y's. Thereafter, we argue that parameterizing the controllers to be linear in nature's y's allows us to encompass linear observation–action policies.

Proof We introduce an augmented partially observed linear system that helps us write the observation–action pairs produced by a linear policy as a linear function of nature's y's. In fact, we will shortly see that, for any sequence of control inputs, the augmented linear system produces the same sequence of obeservations as the original system. Thus, the new system has the same nature's y's as the original system.

For any partially observed system (A, B, C) and perturbation sequence $\mathbf{w}_t$ producing observations $\mathbf{y}_t$, construct a *lifted* system $(\tilde{A}, \tilde{B}, \tilde{C})$ given by

$$\begin{bmatrix} \mathbf{x}_{t+1}^{\text{nat}} \\ \Delta_{t+1} \end{bmatrix} = \begin{bmatrix} A & 0 \\ 0 & A \end{bmatrix} \begin{bmatrix} \mathbf{x}_t^{\text{nat}} \\ \Delta_t \end{bmatrix} + \begin{bmatrix} 0 \\ B \end{bmatrix} \mathbf{u}_t + \begin{bmatrix} I \\ 0 \end{bmatrix} \mathbf{w}_t, \quad \tilde{\mathbf{y}}_t = \begin{bmatrix} C & C \end{bmatrix} \begin{bmatrix} \mathbf{x}_t^{\text{nat}} \\ \Delta_t \end{bmatrix}.$$

Fix any sequence of control inputs $\mathbf{u}_t$. Let $\mathbf{x}_t, \mathbf{y}_t$ be the states and observations produced by the original partially observed system. The equivalents for the augmented system are $[\mathbf{x}_t^{\text{nat}\top}, \Delta_t^\top]^\top, \tilde{\mathbf{y}}_t$, as we have described above. Now we can verify for all times t that $\tilde{\mathbf{y}}_t = \mathbf{y}_t$ and $\mathbf{x}_t = \mathbf{x}_t^{\text{nat}} + \Delta_t$.

Since this new system has the same input-output behavior as the original system, we might equivalently investigate how the execution of a linear controller π affects the augmented system. Substituting $\mathbf{u}_t^\pi = K\mathbf{y}_t^\pi = KC(\mathbf{x}_t^{\text{nat}} + \Delta_t^\pi)$, the state–action sequence produced by a linear policy $\mathbf{u}_t^\pi = K\mathbf{y}_t^\pi$ can be written as

$$\begin{bmatrix} \mathbf{x}_{t+1}^{\text{nat}} \\ \Delta_{t+1}^\pi \end{bmatrix} = \begin{bmatrix} A & 0 \\ BKC & A + BKC \end{bmatrix} \begin{bmatrix} \mathbf{x}_t^{\text{nat}} \\ \Delta_t^\pi \end{bmatrix} + \begin{bmatrix} I \\ 0 \end{bmatrix} \mathbf{w}_t, \quad \begin{bmatrix} \mathbf{y}_t^\pi \\ \mathbf{u}_t^\pi \end{bmatrix} = \begin{bmatrix} C & C \\ KC & KC \end{bmatrix} \begin{bmatrix} \mathbf{x}_t^{\text{nat}} \\ \Delta_t^\pi \end{bmatrix},$$

or using that $\mathbf{y}_t^{\text{nat}} = C\mathbf{x}_t^{\text{nat}}$, equivalently as

$$\Delta_{t+1}^\pi = (A + BKC)\Delta_t^\pi + BK\mathbf{y}_t^{\text{nat}}, \quad \mathbf{y}_t^\pi = C\Delta_t^\pi + \mathbf{y}_t^{\text{nat}}, \quad \mathbf{u}_t^\pi = KC\Delta_t + K\mathbf{y}_t^{\text{nat}}.$$

Now, define $M_i = KC(A + BKC)^{i-1}BK$ for all positive $i \in [h']$ and $M_0 = K$. These parameters specify our DRC. Let $\mathbf{y}_t, \mathbf{u}_t$ be the observation–action sequence produced by this DRC. We then have

$$\begin{aligned}
\mathbf{u}_t^\pi &= K\mathbf{y}_t^{\text{nat}} + KC\Delta_t^\pi \\
&= K\mathbf{y}_t^{\text{nat}} + KCBK\mathbf{y}_{t-1}^{\text{nat}} + KC(A+BKC)\Delta_{t-1}^\pi \\
&= K\mathbf{y}_t^{\text{nat}} + \sum_{i>1} KC(A+BKC)^{i-1}BK\mathbf{y}_{t-i}^{\text{nat}} \\
&= \sum_{i=0}^{h'} M_i \mathbf{y}_{t-i}^{\text{nat}} + \sum_{i>H'} KC(A+BKC)^{i-1}BK\mathbf{y}_{t-i}^{\text{nat}} \\
&= \mathbf{u}_t + Z_t.
\end{aligned}$$

Given γ-stabilizability of the linear controller K and γ-intrinsic stability of the of the system (A, B, C), we have that for some κ, δ such that $\frac{\kappa}{\delta} \leq \gamma$,

$$\begin{aligned}
\|Z_t\| &\leq \sum_{i>h'} \|KC(A+BKC)^{i-1}BK\| \|\mathbf{y}_{t-i}^{\text{nat}}\| \\
&\leq \sum_{i>h'} \|KC(A+BKC)^{i-1}BK\| \left(\sum_{j>1} \|CA^{j-1}\mathbf{w}_{t-i-j}\| \right) \\
&\leq \sum_{i>h'} \|K\|^2 \kappa(1-\delta)^{i-1} \sum_{j>1} \kappa(1-\delta)^j \\
&\leq \frac{\kappa^4}{\delta} \sum_{i>h'}(1-\delta)^{i-1} \leq \frac{\kappa^4}{\delta^2}(1-\delta)^{h'} \leq \frac{\kappa^4}{\delta^2} e^{-\delta h'}.
\end{aligned}$$

Therefore, $\|\mathbf{u}_t - \mathbf{u}_t^\pi\| \leq \varepsilon$ by the appropriate choice of h'. Once again using the *intrinsic* stability of the LTI (A, B, C), we have

$$\|\mathbf{y}_t^\pi - \mathbf{y}_t\| = \left\| \sum_{i=1}^{t} CA^{i-1}B(\mathbf{u}_{t-i}^\pi - \mathbf{u}_{t-i}) \right\|$$
$$\leq \left\| \sum_{i=0}^{t} CA^i B \right\| \max_{i \leq t} \|\mathbf{u}_i^\pi - \mathbf{u}_i\| \leq \frac{\kappa}{\delta}\varepsilon.$$

The approximation in terms of observations and controls implies ε-approximation of the policy class for any Lipschitz cost sequence, concluding the proof. □

12.4 Summary

In this chapter, we extended our study of control to the partially observed setting, where the controller no longer has direct access to the system state but must instead rely on noisy or incomplete observations. This shift fundamentally changes the nature of the control problem, requiring the design of policies that can infer and act on the basis of limited information.

We introduced a hierarchy of policy classes, starting with linear observation–action controllers, which directly map observations to actions, and progressing to generalized linear policies and linear dynamic controllers (LDCs), which incorporate history to improve decision making. This increasing expressiveness highlighted a key limitation: Memoryless policies are insufficient for optimal control, making structured state-dependent controllers necessary.

A major contribution of this chapter was the introduction of disturbance response controllers (DRCs). Instead of explicitly estimating the state of the system, DRCs act directly on a counterfactual signal, "Nature's y's," capturing what the system would have observed in the absence of control inputs. We established that DRCs are a convex relaxation of LDCs, making them computationally attractive while maintaining strong expressive power. This result provides a principled approach to designing policies that are efficient and robust in partially observed environments.

This chapter also set the stage for online nonstochastic control under partial observability, which we will explore in the next chapter. By establishing policy expressiveness results and identifying convex approximations for optimal control, we have built the necessary framework for developing learning-based algorithms that operate effectively in uncertain and adversarial settings.

12.5 Bibliographic Remarks

The study of control in partially observed linear dynamical systems has a long history, with foundational contributions dating back to optimal estimation and control theories. This section provides an overview of key references related to the topics discussed in this chapter, including classical works, expressivity of policy classes, convex relaxations, and connections to online learning.

The problem of control under partial observation was first rigorously formulated in the context of the linear quadratic Gaussian (LQG) regulator, which combines Kalman filtering with optimal control (Anderson and Moore, 1979). The certainty equivalence principle, which states that optimal controllers can be designed assuming a perfect state estimate, was shown to have limitations in more general partially observed settings (Wonham, 1968).

This chapter introduced a hierarchy of policy classes, including linear observation–action controllers, generalized linear controllers, linear dynamic controllers (LDCs), and disturbance response controllers (DRCs). The concept of disturbance response controllers was inspired by parameterizations used in robust control, particularly the Youla–Kucera parameterization (Youla et al., 1976).

Convex relaxations have been a powerful tool for simplifying control design in partially observed LDSs. We have surveyed convex relaxations for learning in dynamical systems in the previous part of this book. Recently, spectral methods for control (Arora et al., 2018) have provided an alternative approach to convex relaxation, utilizing spectral filtering to bypass direct system identification.

The disturbance response parametrization was introduced in Simchowitz et al. (2020) as an analogue of DAC to control partially observed systems. It can be viewed as a succinct state-space representation of the Youla parameterization (Youla et al., 1976). For time-varying dynamics, the power of various policy classes was contrasted in Minasyan et al. (2021).

12.6 Exercises

12.1. Prove Lemma 12.7, stating that the sequence $\{\mathbf{y}_t^{\text{nat}}\}$ can be iteratively computed using only the sequence of observations

$$\mathbf{y}_t^{\text{nat}} = \mathbf{y}_t - C \sum_{i=1}^{t} A^i B \mathbf{u}_{t-i}.$$

Alternatively, the above computation can be carried out recursively as stated below starting with $\mathbf{z}_0 = \mathbf{0}$:

$$\mathbf{z}_{t+1} = A\mathbf{z}_t + B\mathbf{u}_t,$$
$$\mathbf{y}_t^{\text{nat}} = \mathbf{y}_t - C\mathbf{z}_t.$$

12.2. Consider a seemingly more general definition of a partially observed linear dynamical system, given below:

$$\mathbf{x}_{t+1} = A_t \mathbf{x}_t + B_t \mathbf{u}_t + \mathbf{w}_t,$$
$$\mathbf{y}_t = C_t \mathbf{x}_t + \mathbf{v}_t.$$

Construct a linear dynamical system parameterized via matrices (A_t', B_t', C_t') and perturbations $\mathbf{w}_t'$ with a hidden state $\mathbf{z}_t$ of the form

$$\mathbf{x}_{t+1} = A_t' \mathbf{z}_t + B_t' \mathbf{u}_t + \mathbf{w}_t',$$
$$\mathbf{y}_t' = C_t' \mathbf{z}_t,$$

such that for any sequence of $\mathbf{u}_t$, $\mathbf{y}_t = \mathbf{y}_t'$ holds uniformly over time. Additionally, verify that there exists a construction of $\mathbf{z}_t$ where $\mathbf{w}_t'$ are i.i.d. as long as $(\mathbf{w}_t, \mathbf{v}_t)$ are i.i.d. across time steps.

Hint: Consider the choice $\mathbf{w}_t' = D_t' \begin{bmatrix} \mathbf{w}_t \\ \mathbf{v}_t \end{bmatrix}$ for some D_t'. Also, consider using a hidden state $\mathbf{z}_t$ with a larger number of coordinates than $\mathbf{x}_t$.

12.3.

(i) Prove Lemma 12.3.
(ii) Extend the lemma and prove it for time-varying linear dynamical systems.

12.4. Prove Lemma 12.5.

12.5. For any γ-stable partially observed system, prove that any DRC policy in $\Pi_{H,\gamma}^D$ is γ^2-stabilizing.

12.6. Construct a partially observed linear dynamical system subject to Gaussian i.i.d. perturbations with quadratic costs such that it is simultaneously true that (a) there exists a genearlized linear policy that incurs zero (or at most constant) aggregate cost and (b) any linear observation–action policy incurs $\Omega(T)$ aggregate cost when run for T time steps.

Hint: Consider a three-dimensional hidden state so that u_t can be chosen as a function of y_{t-1} to ensure zero cost at each time step.

13
Online Nonstochastic Control with Partial Observation

In the previous chapter, we introduced disturbance response controllers (DRCs) as a structured class of policies for controlling partially observed linear dynamical systems (PO-LDSs). Unlike classical observer-based controllers, which rely on explicit state estimation, DRCs act directly on Nature's y's, the system's counterfactual response in the absence of control. Although DRCs offer a powerful alternative to state-based policies, we have not yet integrated them into an online learning algorithm, leaving open the question of how to adapt them dynamically in uncertain environments.

This chapter takes the next step by introducing an online learning algorithm for optimizing DRCs in real time. We introduce the gradient response controller (GRC), which efficiently updates DRC policies using online convex optimization techniques. The GRC algorithm enables controllers to:

- Learn from past disturbances and actions, refining policy parameters without requiring full system identification.
- Handle adversarial disturbances, extending the policy regret framework to the partially observed setting.
- Optimize control decisions dynamically, adapting to system behavior as new observations arrive.

Unlike previous approaches that relied on fixed policy classes, the GRC provides an adaptive control mechanism that bridges the gap between structured controllers and learning-based methods. A key focus of this chapter is to extend policy regret minimization from the fully observed case to partial observability, introducing new challenges and solutions in learning-based control.

To simplify the theoretical analysis, we assume system stability, allowing us to focus on the core learning problem without additional stabilization constraints. However, we discuss possible extensions to stabilizable systems in the Bibliographic Remarks section.

13.1 The Gradient Response Controller

Partial observation requires modification to the gradient perturbation controller we studied in the previous chapter. The gradient response controller algorithm is given as Algorithm 13.1, for linear time-invariant dynamical systems with partial observation. Similarly to GPC, the main idea is to learn the parameterization of a disturbance response control policy using known online convex optimization algorithms, such as online gradient descent. We henceforth prove that this parameterization is convex, allowing us to prove the main regret bound.

In Algorithm 13.1 we denote by $\mathbf{y}_t(M_{1:h})$ the observation arising from playing the DRC policy $M_{1:h}$ from the beginning of time, and the same for $\mathbf{u}_t(M_{1:h})$. Although it may be impossible to determine $\mathbf{w}_t$, especially if C has fewer rows than columns, from the observations $\mathbf{y}_t$, $\mathbf{y}_t(M_{1:h})$ can be determined from $\mathbf{y}_t^{\text{nat}}$ alone by the application of Lemma 12.7, by

$$\mathbf{y}_t(M_{1:h}) = \mathbf{y}_t^{\text{nat}} + C \sum_{i=1}^{t} A^i B \mathbf{u}_{t-i}(M_{1:h}).$$

Algorithm 13.1 Gradient Response Controller (GRC)

1: Input: system $\{A, B, C\}$, h, η, initialization $M_{0:h}^1 \in \mathcal{K}$.
2: **for** $t = 1 \ldots T$ **do**
3:   Use Control $\mathbf{u}_t = \sum_{i=0}^{h} M_i^t \mathbf{y}_{t-i}^{\text{nat}}$
4:   Observe $\mathbf{y}_{t+1}$, compute for $t+1$:

$$\mathbf{y}_t^{\text{nat}} = \mathbf{y}_t - C \sum_{i=0}^{t} A^i B \mathbf{u}_{t-i}.$$

5:   Construct loss $\ell_t(M_{0:h}) = c_t(\mathbf{y}_t(M_{0:h}), \mathbf{u}_t(M_{0:h}))$
6:   Update $M_{0:h}^{t+1} \leftarrow \prod_{\mathcal{K}} \left[M_{0:h}^t - \eta \nabla \ell_t(M_{0:h}^t) \right]$
7: **end for**

The GRC algorithm comes with a finite-time performance guarantee: It guarantees vanishing worst-case regret versus the best disturbance response control policy in hindsight. Following the relationship between disturbance response control and other policies studied in Section 12.3.1, this implies vanishing regret versus linear dynamical controllers. The formal statement is given as follows.

Theorem 13.1 *Assuming that*

(a) *The costs c_t are convex, bounded, and have bounded gradients with respect to the arguments $\mathbf{y}_t$ and $\mathbf{u}_t$.*
(b) *The diameter of the constraint set for the parameters $M_{0:h}$ is bounded by D.*
(c) *The matrices $\{A, B, C\}$ have bounded ℓ_2 norms.*
(d) *The linear dynamical system $\{A, B, C\}$ is γ-stable.*

Then the GRC (Algorithm 13.1) ensures that

$$\max_{\mathbf{w}_{1:T}:\|\mathbf{w}_t\|\leq 1}\left(\sum_{t=1}^{T}c_t(\mathbf{y}_t,\mathbf{u}_t) - \min_{\pi\in\Pi^R}\sum_{t=1}^{T}c_t(\mathbf{y}_t^\pi,\mathbf{u}_t^\pi)\right) \leq O(LGD\gamma^3 h\sqrt{T}).$$

Furthermore, the time complexity of each loop of the algorithm is polynomial in the number of system parameters and logarithmic in T.

The proof is largely analogous to the regret bound of the GPC presented in Theorem 7.2.

Proof According to Theorem 6.1, the online gradient descent algorithm for convex loss functions ensures that the regret is bounded as a sublinear function of the number of iterations. However, to directly apply the OGD algorithm to our setting, we need to make sure that the following conditions hold true:

- The loss function should be a convex function in the variables $M_{0:h}$.
- We must ensure that optimizing the loss function ℓ_t as a function of parameters $M_{0:h}$ is similar to optimizing the cost c_t as a function of observation and control.

First, we show that the loss function is convex with respect to the variables $M_{0:h}$. This follows since the states and the controls are linear transformations of the variables.

Lemma 13.2 *The loss functions ℓ_t are convex in the variables $M_{0:h}$.*

Proof Using $\mathbf{y}_t$ to denote $\mathbf{y}_t(M_{0:h})$ (and similarly for $\mathbf{u}_t$), the loss function ℓ_t is given by

$$\ell_t(M_{0:h}) = c_t(\mathbf{y}_t, \mathbf{u}_t).$$

13.1 The Gradient Response Controller

Since the cost c_t is a convex function with respect to its arguments, we simply need to show that $\mathbf{y}_t$ and $\mathbf{u}_t$ depend linearly on $M_{0:h}$. Using Lemma 12.7,

$$\mathbf{u}_t = \sum_{i=0}^{h} M_i \mathbf{y}_{t-i}^{\text{nat}},$$

$$\mathbf{y}_t = \mathbf{y}_t^{\text{nat}} + C \sum_{i=1}^{t} A^i B \mathbf{u}_{t-i}$$

$$= \mathbf{y}_t^{\text{nat}} + C \sum_{i=1}^{t} A^i B \sum_{j=0}^{h} M_j \mathbf{y}_{t-i-j}^{\text{nat}},$$

which is a linear function of the variables. Thus, we have shown that $\mathbf{y}_t$ and $\mathbf{u}_t$ are linear transformations in $M_{0:h}$ and, hence, the loss function is convex in $M_{0:h}$. □

Next, we use the following lemma.

Lemma 13.3 *For all t, $|l_t(M_{0:h}^t) - c_t(\mathbf{y}_t, \mathbf{u}_t)| \leq \frac{L\gamma^3 ChD}{\sqrt{T}}$.*

The regret bound of GRC can be derived from these lemmas as follows. Note that G is an upper bound on the gradients of ℓ_t as a function of the parameters $M_{0:h}$, while L is the Lipschitz constant of c_t as a function of the state and control.

$$\sum_{t=1}^{T} c_t(\mathbf{y}_t, \mathbf{u}_t) - \min_{\pi \in \Pi^R} \sum_{t=1}^{T} c_t(\mathbf{y}_t^\pi, \mathbf{u}_t^\pi)$$

$$\leq \sum_{t=1}^{T} l_t(M_{1:h}^t) - \min_{\pi \in \Pi^R} l_t(M_{1:h}^\pi) + T \times \frac{L\gamma^3 ChD}{\sqrt{T}} \quad \text{Lemma 13.3}$$

$$\leq 2GD\sqrt{T} + \sqrt{T} L\gamma^3 ChD \quad \text{Theorem 6.1}$$

$$\leq O(GDL\gamma^3 h\sqrt{T}).$$

□

13.1.1 Finishing Up the Regret Bound

Proof of Lemma 13.3 We first note that by the choice of step size by the online gradient descent (Algorithm 6.1), as per Theorem 6.1, we have $\eta = \frac{D}{G\sqrt{T}}$. Thus,

$$\|M_{0:h}^t - M_{0:h}^{t-i}\| \leq \sum_{s=t-i+1}^{t} \|M_{0:h}^s - M_{0:h}^{s-1}\| \leq i\eta G = \frac{iD}{\sqrt{T}}.$$

Now, we use the fact that the iterates of OGD $M_{1:h}^t$ move slowly to establish that $\mathbf{y}_t$ and $\mathbf{y}_t(M_{0:h}^t)$ are close in value. Before that, by definition, we have $\mathbf{u}_t(M_{0:h}^t) = \mathbf{u}_t$ since the sequence $\mathbf{y}_t^{\text{nat}}$ is determined independently of the choice of the policy that is being executed. Due to the γ-stability of the underlying system, for any $i > 0$, $\|CA^i B\| \leq \kappa(1-\delta)^i$ for some $\kappa > 0, \gamma < 1$. Using Lemma 12.7, we have for some $C > 0$ that

$$\|\mathbf{y}_t(M_{0:h}^t) - \mathbf{y}_t\| = \left\| C \sum_{i=1}^{t} [A_{i-1} B(\mathbf{u}_{t-i}(M_{0:h}^t) - \mathbf{u}_{t-i})] \right\|$$

$$\leq \sum_{i=1}^{t} \|CA^{i-1} B\| \|(\mathbf{u}_{t-i}(M_{0:h}^t) - \mathbf{u}_{t-i})\|$$

$$\leq Ch\kappa \sum_{i=1}^{t} (1-\delta)^i \frac{i}{\sqrt{T}} \max_j \|\mathbf{y}_j^{\text{nat}}\|$$

$$\leq \frac{Ch\kappa D}{\delta^2 \sqrt{T}} \max_j \|\mathbf{y}_j^{\text{nat}}\| \leq \frac{ChD\gamma^3}{\sqrt{T}}.$$

Here we used the fact that $\|\mathbf{y}_t^{\text{nat}}\| \leq \gamma$, which is left as an exercise.

The control depends only on the parameters of the current iteration, and we have $\mathbf{u}_t = \mathbf{u}_t(M_{1:h}^t)$. By definition, $l_t(M_{0:h}^t) = c_t(\mathbf{y}_t(M_{0:h}^t), \mathbf{u}_t(M_{0:h}^t))$. Thus, we have

$$l_t(M_{0:h}^t) - c_t(\mathbf{y}_t, \mathbf{u}_t) = c_t(\mathbf{y}_t(M_{0:h}^t), \mathbf{u}_t(M_{0:h}^t)) - c_t(\mathbf{y}_t, \mathbf{u}_t)$$

$$\leq L \cdot \|\mathbf{y}_t(M_{1:h}^t) - \mathbf{y}_t\| \leq \frac{L\gamma^3 ChD}{\sqrt{T}}.$$

$\square$

13.2 Extension to Time-Varying Systems

The GRC algorithm can be extended to time-varying dynamics, as we explain in this section. The linear dynamical system at time t is denoted by the system matrices A_t, B_t, C_t, and we assume that they are known to the controller. We further assume that these dynamical systems are all γ-stable. The algorithm is formally spelled out in Algorithm 13.2.

It is possible to prove a similar regret bound to the one we proved for linear time-invariant systems. An important step is to note that the observation $\mathbf{y}_t(M_{1:h})$ can be determined from $\mathbf{y}_t^{\text{nat}}$ alone by the application of Lemma 12.7, by

$$\mathbf{y}_t(M_{1:h}) = \mathbf{y}_t^{\text{nat}} + C_t \sum_{i=1}^{t} \left[\prod_{j=1}^{i-1} A_{t-j} \right] B_{t-i} \mathbf{u}_{t-i}(M_{1:h}).$$

13.3 Conclusion

Algorithm 13.2 Gradient Response Controller (GRC)
1: Input: sequence $\{A_t, B_t, C_t\}$, h, η, initialization $M_{0:h}^1 \in \mathcal{K}$.
2: **for** $t = 1 \ldots T$ **do**
3:     Use Control $\mathbf{u}_t = \sum_{i=0}^{h} M_i^t \mathbf{y}_{t-i}^{\text{nat}}$
4:     Observe $\mathbf{y}_{t+1}$, compute for $t+1$:

$$\mathbf{y}_t^{\text{nat}} = \mathbf{y}_t - C_t \sum_{i=0}^{t} \left[\prod_{j=0}^{i} A_{t-j} \right] B_{t-i} \mathbf{u}_{t-i}.$$

5:     Construct loss $\ell_t(M_{0:h}) = c_t(\mathbf{y}_t(M_{0:h}), \mathbf{u}_t(M_{0:h}))$
      Update $M_{0:h}^{t+1} \leftarrow \prod_\mathcal{K} \left[M_{0:h}^t - \eta \nabla \ell_t(M_{0:h}^t) \right]$
6: **end for**

13.3 Conclusion

In this chapter, we extended the framework of online nonstochastic control to the partially observed setting, where the controller must make decisions without direct access to the system state. Unlike the fully observed case, where state-feedback policies can be applied directly, partial observability introduces fundamental challenges, requiring policies that infer system behavior from past observations and actions.

Building on the disturbance response controller (DRC) policies introduced in the previous chapter, we developed the gradient response controller (GRC), an online learning algorithm that optimizes DRC policies in real time.

The results in this chapter demonstrate that adaptive control under partial observability is possible without explicit state estimation, bridging the gap between classical observer-based controllers and modern learning-based methods. Using regret minimization techniques, we provide a robust and flexible approach that applies even in adversarial and uncertain environments.

13.4 Bibliographic Remarks

Online nonstochastic control with partial observability was studied in Simchowitz et al. (2020), who defined the gradient response controller as well as the policy class of disturbance response controllers.

The setting was further extended to time-varying linear dynamics in Minasyan et al. (2021), where the expressive power of different classes of policy was also studied.

The analogue of online nonstochastic control in the stochastic noise setting is called the linear quadraic Gaussian problem, and it is further limited by assuming quadratic noise. This classical problem in optimal control was proposed in Kalman's work (Kalman, 1960). Regret bounds for controlling unknown partially-observed systems were studied in Lale et al. (2020b,c) and Mania et al. (2019). Lale et al. (2020a), which use DRC policies even in the context of stochastic control, obtain rate-optimal logarithmic bounds for the LQG problem.

In Sun et al. (2023), the task of controlling a partially observed system subject to semi-adversarial noise under bandit feedback was considered. The tight rates for quadratic control with bandit feedback, in the fully adversarial model, were obtained in Suggala et al. (2024), and more general cost functions in the bandit setting were considered in Sun and Lu (2024).

13.5 Exercises

13.1. In this exercise, we consider the computation of nature's y's for time-varying systems. Let

$$\mathbf{x}_{t+1}^{\text{nat}} = A_t \mathbf{x}_t^{\text{nat}} + \mathbf{w}_t = \sum_{i=0}^{t} \left[\prod_{j=0}^{i-1} A_{t-j} \right] \mathbf{w}_{t-i} = \Phi_t \mathbf{w}_{1:t}.$$

$$\mathbf{y}_{t+1}^{\text{nat}} = C_{t+1} \mathbf{x}_{t+1}^{\text{nat}} = C_{t+1} \Phi_t \mathbf{w}_{1:t}.$$

Here Φ_t is the linear operator such that $\Phi_t(t-i) = \prod_{j=0}^{i-1} A_{t-j}$. Prove that the sequence $\{\mathbf{y}_t^{\text{nat}}\}$ can be iteratively computed using only the sequence of observations

$$\mathbf{y}_t^{\text{nat}} = \mathbf{y}_t - C_t \sum_{i=1}^{t} \left[\prod_{j=1}^{i-1} A_{t-j} \right] B_{t-i} \mathbf{u}_{t-i}.$$

Alternatively, the above computation can be carried out recursively as stated below starting with $\mathbf{z}_0 = \mathbf{0}$:

$$\mathbf{z}_{t+1} = A_t \mathbf{z}_t + B_t \mathbf{u}_t,$$
$$\mathbf{y}_t^{\text{nat}} = \mathbf{y}_t - C_t \mathbf{z}_t.$$

13.2. Consider a partially observed linear dynamical system that is γ-stable in the absence of control inputs. Prove that when a DRC policy $\pi \in \Pi_{H,\gamma}^R$ is executed on such a dynamical system, the resulting system is γ^2-stable.

13.3. Prove that the optimal controller that minimizes the aggregate cost for a partially observed linear dynamical system assuming that the perturbations are stochastic is a linear dynamical controller.

13.5 Exercises

13.4. Prove that the regret of the GRC controller for time-varying linear dynamical systems is bounded by $O(\sqrt{T})$.
Hint: Follow the proof of Theorem 13.1 and modify as required.

References

Abbasi-Yadkori, Yasin, and Szepesvári, Csaba. 2011. Regret bounds for the adaptive control of linear quadratic systems. Pages 1–26 of: *Proceedings of the 24th Annual Conference on Learning Theory, Budapest, Hungary, 2011*.

Abbasi-Yadkori, Yasin, Bartlett, Peter, and Kanade, Varun. 2014. Tracking adversarial targets. Pages 369–377 of: *International Conference on Machine Learning, Beijing, China, 2014*.

Adam, David. 2020. Special report: The simulations driving the world's response to COVID-19. *Nature*, **580**(7802), 316–319.

Agarwal, Alekh, Jiang, Nan, Kakade, Sham M, and Sun, Wen. 2019a. Lectures on the theory of reinforcement learning. *Available online at rltheorybook.github.io*.

Agarwal, Naman, Hazan, Elad, and Singh, Karan. 2019b. Logarithmic regret for online control. Pages 10175–10184 of: *Advances in Neural Information Processing Systems, Vancouver, British Columbia, Canada, 2019*.

Agarwal, Naman, Bullins, Brian, Hazan, Elad, Kakade, Sham, and Singh, Karan. 2019c. Online control with adversarial disturbances. Pages 111–119 of: *International Conference on Machine Learning, Los Angeles, California, United States, 2019*.

Agarwal, Naman, Hazan, Elad, Majumdar, Anirudha, and Singh, Karan. 2021. A regret minimization approach to iterative learning control. Pages 100–109 of: *International Conference on Machine Learning, Vienna, Austria, 2021*.

Agarwal, Naman, Suo, Daniel, Chen, Xinyi, and Hazan, Elad. 2023. Spectral state space models. *arXiv preprint arXiv:2312.06837*.

Agarwal, Naman, Chen, Xinyi, Dogariu, Evan, Feinberg, Vlad, Suo, Daniel, Bartlett, Peter, and Hazan, Elad. 2024. FutureFill: Fast generation from convolutional sequence models. *arXiv preprint arXiv:2410.03766*.

Ahmadi, Amir Ali. 2012. On the difficulty of deciding asymptotic stability of cubic homogeneous vector fields. Pages 3334–3339 of: *2012 American Control Conference (ACC), Montréal, Canada, 2012*.

Ahmadi, Amir Ali. 2016. *ORF 363: Computing and Optimization*. http://aaa.princeton.edu/orf363. [Online; accessed August 1, 2022].

Anava, Oren, Hazan, Elad, and Mannor, Shie. 2015. Online learning for adversaries with memory: Price of past mistakes. *Advances in Neural Information Processing Systems, Montréal, Quebec, Canada, 2015*.

References

Anderson, BDO, and Moore, John B. 1991. Kalman filtering: Whence, what and whither? *Mathematical system theory: The influence of RE Kalman*. Springer.

Anderson, BDO, and Moore, John B. 1979. *Optimal filtering*. Dover Publications.

Anderson, James, Doyle, John C, Low, Steven H, and Matni, Nikolai. 2019. System level synthesis. *Annual Reviews in Control*, **47**, 364–393.

Arora, Raman, Dekel, Ofer, and Tewari, Ambuj. 2012. Online bandit learning against an adaptive adversary: From regret to policy regret. *arXiv preprint arXiv:1206.6400*.

Arora, Sanjeev. 2020. Personal communications.

Arora, Sanjeev, Hazan, Elad, Lee, Holden, Singh, Karan, Zhang, Cyril, and Zhang, Yi. 2018. Towards Provable Control for Unknown Linear Dynamical Systems. *International Conference on Learning Representations (Workshop Track), Vancouver, Canada, 2018*.

Åström, Karl Johan, and Hägglund, Tore. 1995. *PID controllers: Theory, design, and tuning*. Vol. 2. Instrument society of America Research Triangle Park, NC.

Baby, Dheeraj, and Wang, Yu-Xiang. 2022. Optimal dynamic regret in LQR control. *Advances in Neural Information Processing Systems, New Orleans, Louisiana, United States, 2022*, 24879–24892.

Bakshi, Ainesh, Liu, Allen, Moitra, Ankur, and Yau, Morris. 2023a. A new approach to learning linear dynamical systems. Pages 335–348 of: *Proceedings of the 55th Annual ACM Symposium on Theory of Computing, Orlando, Florida, United States, 2023*.

Bakshi, Ainesh, Liu, Allen, Moitra, Ankur, and Yau, Morris. 2023b. Tensor decompositions meet control theory: Learning general mixtures of linear dynamical systems. Pages 1549–1563 of: *International Conference on Machine Learning, Honolulu, Hawaii, United States, 2023*.

Başar, Tamer, and Bernhard, Pierre. 2008. *H-infinity optimal control and related minimax design problems: A dynamic game approach*. Springer Science & Business Media.

Beckermann, Bernhard, and Townsend, Alex. 2017. On the singular values of matrices with displacement structure. *SIAM Journal on Matrix Analysis and Applications*, **38**(4), 1227–1248.

Bennett, S. 1993. Development of the PID controller. *IEEE Control Systems Magazine*, **13**(6), 58–62.

Bertsekas, Dimitri. 2019. *Reinforcement learning and optimal control*. Athena Scientific.

Bertsekas, Dimitri P. 2007. *Dynamic programming and optimal control*. Athena Scientific.

Blondel, Vincent, and Tsitsiklis, John N. 1997. NP-hardness of some linear control design problems. *SIAM Journal on Control and Optimization*, **35**(6), 2118–2127.

Blondel, Vincent D, and Tsitsiklis, John N. 1999. Three problems on the decidability and complexity of stability. Pages 45–51 of: *Open problems in mathematical systems and control theory*. Springer.

Blondel, Vincent D, and Tsitsiklis, John N. 2000. A survey of computational complexity results in systems and control. *Automatica*, **36**(9), 1249–1274.

Boyd, S. 2010. *EE 263: Introduction to Linear Dynamical Systems, lecture notes*. Available online at ee263.stanford.edu.

References

Brahmbhatt, Anand, Buzaglo, Gon, Druchyna, Sofiia, and Hazan, Elad. 2025. *A New Approach to Controlling Linear Dynamical Systems*. arXiv preprint arXiv:2504.03952.

Braverman, Mark, and Yampolsky, Michael. 2008. *Computability of Julia sets*. Springer.

Cassel, Asaf, and Koren, Tomer. 2020. Bandit linear control. *Advances in Neural Information Processing Systems, Vancouver, British Columbia, Canada, 2020*, 8872–8882.

Cassel, Asaf, Cohen, Alon, and Koren, Tomer. 2020. Logarithmic regret for learning linear quadratic regulators efficiently. Pages 1328–1337 of: *International Conference on Machine Learning, Vienna, Austria, 2020*.

Cassel, Asaf, Cohen, Alon, and Koren, Tomer. 2022. Rate-optimal online convex optimization in adaptive linear control. *arXiv preprint arXiv:2206.01426*.

Cesa-Bianchi, Nicolo, and Lugosi, Gábor. 2006. *Prediction, learning, and games*. Cambridge University Press.

Chen, Xinyi, and Hazan, Elad. 2021. Black-box control for linear dynamical systems. Pages 1114–1143 of: *Conference on Learning Theory, Boulder, Colorado, United States, 2021*.

Cohen, Alon, Hasidim, Avinatan, Koren, Tomer, Lazic, Nevena, Mansour, Yishay, and Talwar, Kunal. 2018. Online linear quadratic control. Pages 1029–1038 of: *International Conference on Machine Learning, Stockholm, Sweden, 2018*.

Cohen, Alon, Koren, Tomer, and Mansour, Yishay. 2019. Learning linear-quadratic regulators efficiently with only $\sqrt{T}$ regret. Pages 1300–1309 of: *International Conference on Machine Learning, Los Angeles, California, United States, 2019*.

Cormen, Thomas H, Leiserson, Charles E, Rivest, Ronald L, and Stein, Clifford. 2022. *Introduction to algorithms*. MIT Press.

Dean, Sarah, Mania, Horia, Matni, Nikolai, Recht, Benjamin, and Tu, Stephen. 2018. Regret bounds for robust adaptive control of the linear quadratic regulator. Pages 4188–4197 of: *Advances in Neural Information Processing Systems, Montréal, Quebec, Canada, 2018*.

Dean, Sarah, Mania, Horia, Matni, Nikolai, Recht, Benjamin, and Tu, Stephen. 2020. On the sample complexity of the linear quadratic regulator. *Foundations of Computational Mathematics*, **20**(4), 633–679.

Doyle, John C. 1978. Guaranteed margins for LQG regulators. *IEEE Transactions on Automatic Control*, **23**(4), 756–757.

Doyle, John C, Matni, Nikolai, Wang, Yuh-Shyang, Anderson, James, and Low, Steven. 2017. System level synthesis: A tutorial. Pages 2856–2867 of: *2017 IEEE 56th Annual Conference on Decision and Control (CDC), Melbourne, Australia, 2017*.

Du, Simon, Kakade, Sham, Lee, Jason, Lovett, Shachar, Mahajan, Gaurav, Sun, Wen, and Wang, Ruosong. 2021. Bilinear classes: A structural framework for provable generalization in RL. Pages 2826–2836 of: *International Conference on Machine Learning, Vienna, Austria, 2021*.

Even-Dar, Eyal, Kakade, Sham M, and Mansour, Yishay. 2004. Experts in a Markov decision process. *Advances in neural information processing systems, Vancouver, British Columbia, Canada, 2004*.

Fazel, Maryam, Ge, Rong, Kakade, Sham, and Mesbahi, Mehran. 2018. Global convergence of policy gradient methods for the linear quadratic regulator. Pages

1467–1476 of: *International Conference on Machine Learning, Stockholm, Sweden, 2018*.

Fernández Cara, Enrique, and Zuazua Iriondo, Enrique. 2003. Control Theory: History, mathematical achievements and perspectives. *Boletín de la Sociedad Española de Matemática Aplicada*, **26**, 79–140.

Fiechter, Claude-Nicolas. 1997. PAC adaptive control of linear systems. Pages 72–80 of: *Proceedings of the 10th Annual Conference on Computational Learning Theory, Nashville, Tennessee, United States, 1997*.

Foster, Dylan J., and Simchowitz, Max. 2020. Logarithmic regret for adversarial online control. *arXiv preprint arXiv:2003.00189*.

Furieri, Luca, Zheng, Yang, Papachristodoulou, Antonis, and Kamgarpour, Maryam. 2019. An input–output parametrization of stabilizing controllers: Amidst youla and system level synthesis. *IEEE Control Systems Letters*, **3**(4), 1014–1019.

Ghai, Udaya, Lee, Holden, Singh, Karan, Zhang, Cyril, and Zhang, Yi. 2020. No-regret prediction in marginally stable systems. Pages 1714–1757 of: *Conference on Learning Theory, Graz, Austria, 2020*.

Goel, Gautam, and Hassibi, Babak. 2021. Competitive control. *arXiv preprint arXiv:2107.13657*.

Goel, Gautam, and Wierman, Adam. 2019. An online algorithm for smoothed regression and lqr control. Pages 2504–2513 of: *The 22nd International Conference on Artificial Intelligence and Statistics, Naha, Okinawa, Japan, 2019*.

Goel, Gautam, Lin, Yiheng, Sun, Haoyuan, and Wierman, Adam. 2019. Beyond online balanced descent: An optimal algorithm for smoothed online optimization. *Advances in Neural Information Processing Systems, Vancouver, British Columbia, Canada, 2019*.

Goel, Gautam, Agarwal, Naman, Singh, Karan, and Hazan, Elad. 2023. Best of both worlds in online control: Competitive ratio and policy regret. Pages 1345–1356 of: *Learning for Dynamics and Control Conference, Philadelphia, Pennsylvania, United States, 2023*.

Golowich, Noah, Hazan, Elad, Lu, Zhou, Rohatgi, Dhruv, and Sun, Y Jennifer. 2024. Online control in population dynamics. *arXiv preprint arXiv:2406.01799*.

Gordon, Neil J, Salmond, David J, and Smith, Adrian FM. 1993. Novel approach to nonlinear/non-Gaussian Bayesian state estimation. Pages 107–113 of: *IEEE Proceedings F (Radar and Signal Processing), 1993*, vol. 140.

Gradu, Paula, Hazan, Elad, and Minasyan, Edgar. 2020a. Adaptive regret for control of time-varying dynamics. *arXiv preprint arXiv:2007.04393*.

Gradu, Paula, Hallman, John, and Hazan, Elad. 2020b. Non-stochastic control with bandit feedback. *Advances in Neural Information Processing Systems, Vancouver, British Columbia, Canada, 2020*, 10764–10774.

Hardt, Moritz, Ma, Tengyu, and Recht, Benjamin. 2018. Gradient descent learns linear dynamical systems. *Journal of Machine Learning Research*, **19**(29), 1–44.

Hassibi, Babak, Sayed, Ali H, and Kailath, Thomas. 1996. Linear estimation in Krein spaces. II. Applications. *IEEE Transactions on Automatic Control*, **41**(1), 34–49.

Hazan, Elad. 2016. Introduction to online convex optimization. *Foundations and Trends® in Optimization*, **2**(3–4), 157–325, Now Publishers, Inc.

Hazan, Elad. 2019. Lecture notes: Optimization for machine learning. *arXiv preprint arXiv:1909.03550*.

Hazan, Elad, and Singh, Karan. 2021. ICML tutorial: Online and nonstochastic control. In: *International Conference on Machine Learning, Vienna, Austria, 2021.*

Hazan, Elad, Singh, Karan, and Zhang, Cyril. 2017. Learning linear dynamical systems via spectral filtering. Pages 6702–6712 of: *Advances in Neural Information Processing Systems, Long Beach, California, United States, 2017.*

Hazan, Elad, Lee, Holden, Singh, Karan, Zhang, Cyril, and Zhang, Yi. 2018. Spectral filtering for general linear dynamical systems. Pages 4634–4643 of: *Advances in Neural Information Processing Systems, Montréal, Quebec, Canada, 2018.*

Hazan, Elad, Kakade, Sham, and Singh, Karan. 2020. The nonstochastic control problem. Pages 408–421 of: *Algorithmic Learning Theory, San Diego, California, United States, 2020.*

Hespanha, Joao P. 2018. *Linear systems theory.* Princeton University Press.

Ho, BL, and Kálmán, Rudolf E. 1966. Effective construction of linear state-variable models from input/output functions. *At-Automatisierungstechnik*, **14**(1–12), 545–548.

Ibrahimi, Morteza, Javanmard, Adel, and Roy, Benjamin. 2012. Efficient reinforcement learning for high dimensional linear quadratic systems. *Advances in Neural Information Processing Systems, Stateline, Nevada, United States, 2012.*

Jiang, Nan, Krishnamurthy, Akshay, Agarwal, Alekh, Langford, John, and Schapire, Robert E. 2017. Contextual decision processes with low Bellman rank are pac-learnable. Pages 1704–1713 of: *International Conference on Machine Learning, Sydney, Australia, 2017.*

Jin, Chi, Yang, Zhuoran, Wang, Zhaoran, and Jordan, Michael I. 2020. Provably efficient reinforcement learning with linear function approximation. Pages 2137–2143 of: *Conference on Learning Theory, Graz, Austria, 2020.*

Julier, Simon J, and Uhlmann, Jeffrey K. 2004. Unscented filtering and nonlinear estimation. *Proceedings of the IEEE, 2004*, **92**(3), 401–422.

Kakade, Sham, Krishnamurthy, Akshay, Lowrey, Kendall, Ohnishi, Motoya, and Sun, Wen. 2020. Information theoretic regret bounds for online nonlinear control. *Advances in Neural Information Processing Systems, Vancouver, British Columbia, Canada, 2020*, **33**, 15312–15325.

Kalman, Rudolph Emil. 1960. A new approach to linear filtering and prediction problems. *Journal of Basic Engineering*, **82.1**, 35–45.

Khalil, Hassan K. 2015. *Nonlinear control.* Pearson.

Kozdoba, Mark, Marecek, Jakub, Tchrakian, Tigran, and Mannor, Shie. 2019. On-line learning of linear dynamical systems: Exponential forgetting in kalman filters. Pages 4098–4105 of: *Proceedings of the AAAI Conference on Artificial Intelligence, Honolulu, Hawaii, United States, 2019.*

Lale, Sahin, Azizzadenesheli, Kamyar, Hassibi, Babak, and Anandkumar, Anima. 2020a. Adaptive control and regret minimization in linear quadratic gaussian (LQG) setting. *arXiv preprint arXiv:2003.05999.*

Lale, Sahin, Azizzadenesheli, Kamyar, Hassibi, Babak, and Anandkumar, Anima. 2020b. Logarithmic regret bound in partially observable linear dynamical systems. *arXiv preprint arXiv:2003.11227.*

Lale, Sahin, Azizzadenesheli, Kamyar, Hassibi, Babak, and Anandkumar, Anima. 2020c. Regret minimization in partially observable linear quadratic control. *arXiv preprint arXiv:2002.00082.*

References

Levine, Sergey, and Koltun, Vladlen. 2013. Guided policy search. Pages 1–9 of: *International Conference on Machine Learning, Atlanta, Georgia, United States, 2013*.

Li, Weiwei. 2005. A generalized iterative LQG method for locally-optimal feedback control of constrained nonlinear stochastic systems. Pages 300–306 of: *Proceedings of the American Control Conference, Portland, Oregon, United States, 2005*.

Li, Yingying, Das, Subhro, and Li, Na. 2021. Online optimal control with affine constraints. Pages 8527–8537 of: *Proceedings of the AAAI Conference on Artificial Intelligence, 2021*.

Liu, Y Isabel, Nguyen, Windsor, Devre, Yagiz, Dogariu, Evan, Majumdar, Anirudha, and Hazan, Elad. 2024. Flash stu: Fast spectral transform units. *arXiv preprint arXiv:2409.10489*.

Lorenz, Edward N. 1963. Deterministic nonperiodic flow. *Journal of the Atmospheric Sciences*, **20**(2), 130–141.

Lyapunov, Aleksandr Mikhailovich. 1992. The general problem of the stability of motion. *International Journal of Control*, **55**(3), 531–534.

Mania, Horia, Tu, Stephen, and Recht, Benjamin. 2019. Certainty equivalence is efficient for linear quadratic control. Pages 10154–10164 of: *Advances in Neural Information Processing Systems, Vancouver, British Columbia, Canada, 2019*.

Marsden, Annie, and Hazan, Elad. 2025. Dimension-free regret for learning asymmetric linear dynamical systems. *arXiv preprint arXiv:2502.06545*.

Marsden, Annie, Dogariu, Evan, Agarwal, Naman, Chen, Xinyi, Suo, Daniel, and Hazan, Elad. 2024. Provable length generalization in sequence prediction via spectral filtering. *arXiv preprint arXiv:2411.01035*.

Maxwell, James Clerk. 1868. I. On governors. *Proceedings of the Royal Society of London*, **16**, 270–283.

Minasyan, Edgar, Gradu, Paula, Simchowitz, Max, and Hazan, Elad. 2021. Online control of unknown time-varying dynamical systems. *Advances in Neural Information Processing Systems, 2021*, 15934–15945.

Oymak, Samet, and Ozay, Necmiye. 2019. Non-asymptotic identification of LTI systems from a single trajectory. Pages 5655–5661 of: *American Control Conference (ACC), Philadelphia, Pennsylvania, United States, 2019*.

Plevrakis, Orestis, and Hazan, Elad. 2020. Geometric exploration for online control. *Advances in Neural Information Processing Systems, Vancouver, British Columbia, Canada, 2020*, 7637–7647.

Rockafellar, R Tyrell. 1987. Linear-quadratic programming and optimal control. *SIAM Journal on Control and Optimization*, **25**(3), 781–814.

Russel, PR, and Norvig, SA. 2002. *Artificial Intelligence: A modern approach*. Prentice Hall.

Sarkar, Tuhin, and Rakhlin, Alexander. 2019. Near optimal finite time identification of arbitrary linear dynamical systems. Pages 5610–5618 of: *Proceedings of the 36th International Conference on Machine Learning, Los Angeles, California, United States, 2019*. Proceedings of Machine Learning Research.

Sastry, Shankar, Bodson, Marc, and Bartram, James F. 1990. *Adaptive control: Stability, convergence, and robustness*. Acoustical Society of America.

Shi, Guanya, Lin, Yiheng, Chung, Soon-Jo, Yue, Yisong, and Wierman, Adam. 2020. Online optimization with memory and competitive control. *Advances in Neural Information Processing Systems, Vancouver, British Columbia, Canada, 2020*, 20636–20647.

Silver, David. 2015. *Lecture Notes on Reinforcement Learning*. Lecture Notes.

Simchowitz, Max. 2020. Making non-stochastic control (almost) as easy as stochastic. *Advances in Neural Information Processing Systems, Vancouver, British Columbia, Canada, 2020*, 18318–18329.

Simchowitz, Max, and Foster, Dylan. 2020. Naive exploration is optimal for online LQR. Pages 8937–8948 of: *Proceedings of the 37th International Conference on Machine Learning, Baltimore, Maryland, United States. 2020*. Proceedings of Machine Learning Research.

Simchowitz, Max, Mania, Horia, Tu, Stephen, Jordan, Michael I, and Recht, Benjamin. 2018. Learning without mixing: Towards a sharp analysis of linear system identification. Pages 439–473 of: *Conference On Learning Theory, Stockholm, Sweden, 2018*.

Simchowitz, Max, Boczar, Ross, and Recht, Benjamin. 2019. Learning linear dynamical systems with semi-parametric least squares, Phoenix, Arizona, United States, 2019. Pages 2714–2802 of: *Conference on Learning Theory*.

Simchowitz, Max, Singh, Karan, and Hazan, Elad. 2020. Improper learning for non-stochastic control. Pages 3320–3436 of: *Conference on Learning Theory, Graz, Austria, 2020*.

Singh, Karan. 2022. *The nonstochastic control problem*. PhD thesis, Princeton University.

Slotine, Jean-Jacques E, and Li, Weiping. 1991. *Applied nonlinear control*. Prentice Hall.

Stengel, Robert F. 1994. *Optimal control and estimation*. Courier Corporation.

Strogatz, Steven H. 2014. *Nonlinear dynamics and chaos*. Taylor & Francis Inc.

Suggala, Arun, Sun, Y Jennifer, Netrapalli, Praneeth, and Hazan, Elad. 2024. Second order methods for bandit optimization and control. Pages 4691–4763 of: *The 37th Annual Conference on Learning Theory, Edmonton, Canada, 2024*.

Sun, Y Jennifer, and Lu, Zhou. 2024. Tight rates for bandit control beyond quadratics. *arXiv preprint arXiv:2410.00993*.

Sun, Y Jennifer, Newman, Stephen, and Hazan, Elad. 2023. Optimal rates for bandit nonstochastic control. *Advances in Neural Information Processing Systems, New Orleans, Louisiana, United State, 2023*, 21908–21919.

Suo, Daniel, Zhang, Cyril, Gradu, Paula, Ghai, Udaya, Chen, Xinyi, Minasyan, Edgar, Agarwal, Naman, Singh, Karan, LaChance, Julienne, Zajdel, Tom, et al. 2021. Machine learning for mechanical ventilation control. *arXiv preprint arXiv:2102.06779*.

Sutton, Richard S., and Barto, Andrew G. 2018. *Reinforcement learning: An introduction*. Second edn. MIT Press.

Tedrake, Russ. 2020. Underactuated robotics: Algorithms for walking, running, swimming, flying, and manipulation *(Course Notes for MIT 6.832)*.

Tropp, Joel A. 2012. User-friendly tail bounds for sums of random matrices. *Foundations of Computational Mathematics*, **12**, 389–434.

Tu, Stephen. 2018. H-infinity optimal control via dynamic games. *Available online at stephentu.github.io.*

Tucker, Warwick. 1999. The Lorenz attractor exists. *Comptes Rendus de l'Académie des Sciences-Series I-Mathematics,* **328**(12), 1197–1202.

Urrea, Claudio, and Agramonte, Rayko. 2021. Kalman filter: Historical overview and review of its use in robotics 60 years after its creation. *Journal of Sensors,* **2021**(1), 9674015.

Wiener, Norbert. 1949. *Extrapolation, interpolation, and smoothing of stationary time series: With engineering applications.* MIT Press.

Wonham, Walter Murray. 1968. On the separation theorem of stochastic control. *SIAM Journal on Control,* **6**(2), 312–326.

Youla, Dante, Jabr, Hamid, and Bongiorno, Jr. 1976. Modern Wiener-Hopf design of optimal controllers–Part II: The multivariable case. *IEEE Transactions on Automatic Control,* **21**(3), 319–338.

Zhang, Runyu Cathy, Zheng, Yang, Li, Weiyu, and Li, Na. 2023. On the relationship of optimal state feedback and disturbance response controllers. *IFAC-PapersOnLine,* **56**(2), 7503–7508.

Zhao, Peng, Wang, Yu-Xiang, and Zhou, Zhi-Hua. 2022. Non-stationary online learning with memory and non-stochastic control. Pages 2101–2133 of: *International Conference on Artificial Intelligence and Statistics, 2022.*

Zheng, Yang, Furieri, Luca, Papachristodoulou, Antonis, Li, Na, and Kamgarpour, Maryam. 2020. On the equivalence of youla, system-level, and input–output parameterizations. *IEEE Transactions on Automatic Control,* **66**(1), 413–420.

Zhou, Kemin, Doyle, John C, and Glover, Keither. 1996a. Robust and optimal control. *Control Engineering Practice,* **4**(8), 1189–1190.

Zhou, Kemin, Doyle, John C, and Glover, Keith. 1996b. *Robust and optimal control.* Prentice-Hall, Inc.

Index

bang-bang control, 7
Bellman equation, 34
   average reward case, 35
   finite time case, 35

computational complexity, 21
control theory
   history, 4
   optimal control, 8, 50
   robust control, 9, 55

disturbance action controller, 65
   approximating linear policies, 66
disturbance response controller, 131
   approximating other policy classes, 133
   nature's y's, 131
double integrator, 14
dynamical systems
   controllable, 19
   definition, 5
   equilibrium, 18
   examples
      aircraft dynamics, 15
      double integrator, 14
      medical ventilator, 13
      pendulum, 15
      SIR model, 17
   stabilizable, 19

generic control problem, 5
gradient perturbation controller, 75
   regret bound, 76
   simplified, 11
Gradient Response Controller
   regret bound, 140
   time varying systems, 142

Hankel matrix, 115
   eigen-spectrum, 116
H_∞ control, 55

Julia set, 20

Kalman controllability matrix, 44
Kalman filter, 102
   as a linear predictor, 108
   Bayes optimality, 108
   infinite horizon, 107
   optimal linear estimator, 104
Kalman observability matrix, 104

lifting, 129, 133
linear dynamical systems
   γ-stabilizability, 46
   controllability, 44
   definition, 40
   learning, 82
   LTI or time-invariant, 42
   partially observed, 103, 127
   prediction, 93
   stabilizability, 41
      as a semi-definite program, 43
   strong controllability, 47
   unknown system, 82, 93
linear policies, 65
linear quadratic r.egulator, 50
linear quadratic regulator
   infinite horizon, 53
   optimal policy, 52
   Riccati equation, 54
linearization, 41
Lorenz attractor, 20

Markov chain, 28
  ergodic, 30
  reversible, 30
  stationary distribution, 29
Markov decision process, 28
  definition, 32
  Markov reward process, 31
  optimal policy, 34
  Q function, 33
  value function, 32
matrix Azuma inequality, 85
method of moments, 84

nature's y's, 131
nonstochastic control, 10, 73
  definition, 74
  partially observed systems, 139
  system identification, 84

online convex optimization, 60
  online gradient descent, 60
  with memory, 78, 141
online gradient descent, 60

partially observable systems, 103, 127
PID control, 8
policy classes
  ε-approximation, 63
  disturbance action controller, 65
  disturbance response controllers, 131
  generalized linear observation controllers, 129
  linear dynamical controllers, 130
    via generalized linear observation controllers, 131
  linear observation controllers, 128
  linear policies, 65

prediction, 94
  regret bound, 98
prediction rules, 95
  linear dynamical predictors, 96
    via linear predictors, 97
  linear predictors, 96
    learning, 98
  spectral predictors, 116
    approximating linear predictors, 117

regret
  in online convex optimzation, 60
  policy regret, 10
reinforcement learning, 25
  as a linear program, 36
  exploration, 87
  value iteration, 36
Riccati equation, 54, 107
  as a semi-definite program, 54

Schur complement, 43
semi-definite programming, 43, 54
spectral filtering, 113
  filters, 121
  in higher dimensions, 120
  in one dimension, 114
  regret bound, 120
spectral radius, 42
stability
  γ-stability, 19
  BIBO stability, 19
  Lyapunov, 43
  NP-hardness, 21
system identification, 82

value iteration, 36

For EU product safety concerns, contact us at Calle de José Abascal, 56–1°,
28003 Madrid, Spain or eugpsr@cambridge.org.

www.ingramcontent.com/pod-product-compliance
Lightning Source LLC
LaVergne TN
LVHW022003060526
838200LV00003B/71